Th
Scien

A Student Reference Guide

Academic Vocabulary Builders

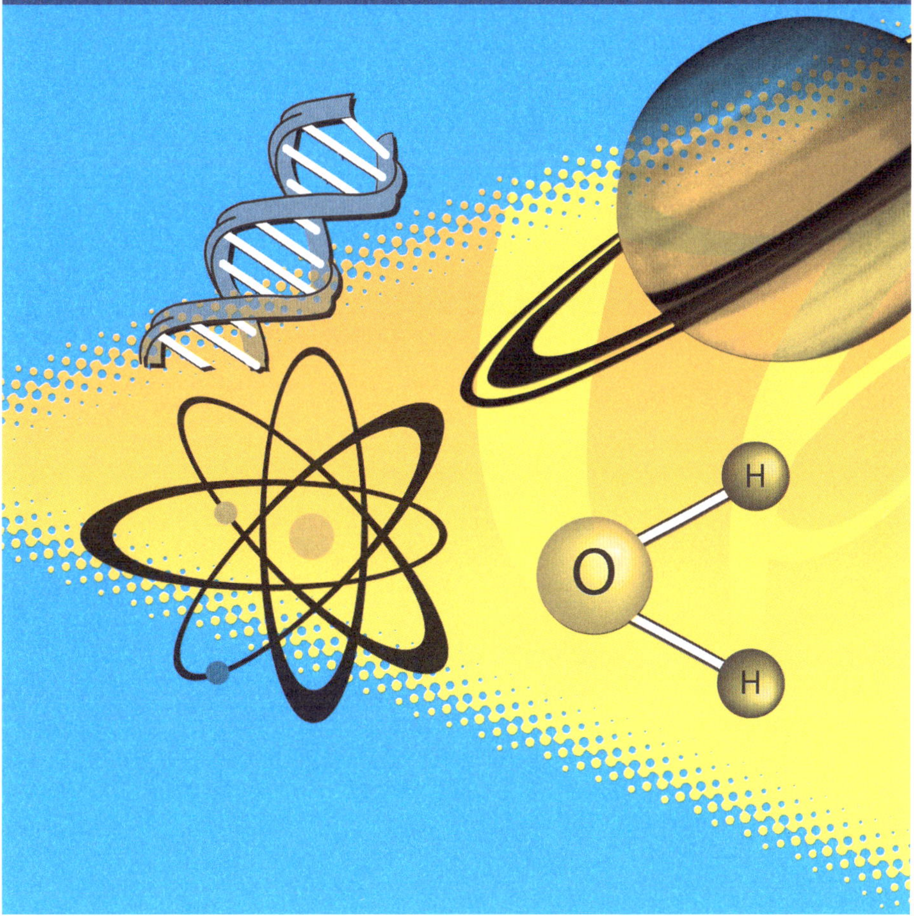

Academic Vocabulary Builders are published by Red Brick Learning
7825 Telegraph Road, Bloomington, Minnesota 55438
http://www.redbricklearning.com

Copyright © 2009 Red Brick Learning.
All rights reserved. No part of this book may be reproduced without written permission from the publisher. The publisher takes no responsibility for the use of any of the materials or methods described in this book, nor for the products thereof.
Printed in the United States of America

Library of Congress Cataloging-in-Publication Data
The essential science glossary III: a student reference guide.
   p. cm. — (Academic vocabulary builders)
   Includes index.
   Summary: "The Level III glossary covers essential content terms in the key subject area of science for high school level students" — Provided by publisher.
   ISBN-13: 978-1-4296-2728-3 (softcover)
   ISBN-10: 1-4296-2728-X (softcover)
   1. Science — Study and teaching (Secondary) 2. Science — Terminology. I. Red Brick Learning (Firm) II. Title: Essential science glossary three. III. Title: Essential science glossary 3.
Q181.E72 2009
500 — dc22
                                                2008021076

Cover Design
Ted Williams

Design and Illustration
Sasha Blanton and Will Hester
SGB Design Solutions

1 2 3 4 5 6 13 12 11 10 09 08

Table of Contents

General Science Terms   1

Biology   12

Chemistry   53

Physical Science   77

Earth and Space   99

Periodic Table of the Elements   128

Academic Vocabulary Builders

About this book:

This book is to help you learn essential words you will need to understand to do well on state tests. These essential words will also help you to do well in school.

There are about 500 Science words and definitions in the book. They are listed in alphabetical order under five main topics.

Here is a sample word with its features:

Easy to read definitions

Pictures to help understanding

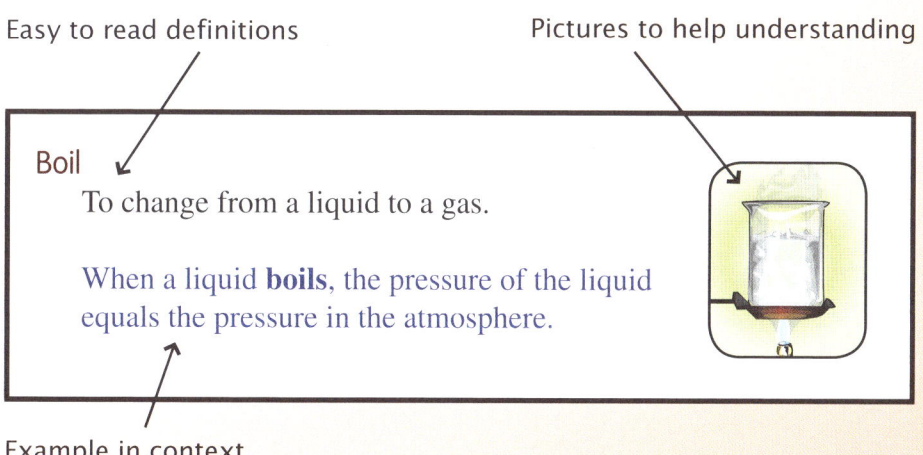

Boil

To change from a liquid to a gas.

When a liquid **boils**, the pressure of the liquid equals the pressure in the atmosphere.

Example in context

v

General Science Terms

Boil
To change from a liquid to a gas.

When a liquid **boils**, the pressure of the liquid equals the pressure in the atmosphere.

Combustion
The chemical change that happens when a substance combines with oxygen to produce heat and light.

Combustion is also known as *burning*.

Compound
A combination of elements that are bonded together, have specific properties, and have a definite composition.

Sucrose, or table sugar, is an example of a **compound**. It is made of a specific number of carbon, hydrogen, and oxygen atoms that are bonded together.

Conclusion
What you learn from doing an experiment.

Ms. Cohen's class studied the effects of sunlight on colored paper. Their **conclusion** was that sunlight turns colored paper white.

1

General Science Terms

Condensation
Turning from a gas to a liquid.

Condensation is the opposite of evaporation.

Conserve
To use something wisely; to not waste something.

Students can **conserve** fuel by walking to school instead of driving a car.

Constant
In an experiment, something that does not change.

If a student does an experiment to learn which plant food works best for roses, a **constant** might be the amount of water each plant gets. The amount of water given should not change for each plant.

Control
In an experiment, a subject that does not have a variable added or taken away.

We compared colored paper that had been exposed to sunlight to a **control**: paper that had not been left in the sun.

Cycle
A series of events in which the last step leads back to the first step.

Cycles happen over and over again.

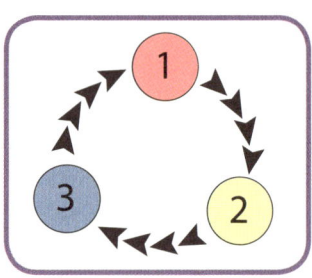

General Science Terms

Data (singular: datum)
Pieces of information.

Scientists gather **data** from the world around them.

Density
A measure of how tightly matter is packed.
Density = mass ÷ volume.

A rock has greater **density** than a foam ball of the same size.

Energy
The ability to do an action.

You need **energy** for a lamp to turn on. This **energy** can include the **energy** your body uses by walking over and flipping the switch as well as the electrical **energy** that comes up through the wires into the lamp.

Equilibrium
The state in which varying, shifting, or opposing forces are in balance.

If you mix fresh water with salt water, the two liquids will reach **equilibrium** when there is an equal amount of salt throughout the mixture.

General Science Terms

Evaporation
Particles that leave a substance as gases when the substance is not boiling.

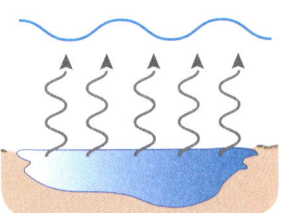

Evaporation is the opposite of condensation.

Evidence
A sign that something is probably true.

Kane thinks that bees like red flowers more than yellow flowers. **Evidence** such as more bees on red flowers might prove that this is true.

Freeze
To change from a liquid to a solid by removing heat.

Gas
Any substance with no shape or volume.

The air we breathe is made up of many **gases**, including oxygen and nitrogen.

General Science Terms

Hypothesis
An educated guess about what will happen.

Scientists can use an experiment to test their **hypothesis** about weathering.

Light
The part of the electromagnetic spectrum that we can see.

Light comes from many sources, including lamps, flashlights, and the Sun.

Liquid
Any substance with volume but no shape.

Water, oil, and honey are all **liquids**.

Magnitude
A measurement that is shown as a number and a standard unit of measurement.

If you measure the length of your pencil, the number might be 6 and the standard unit of measurement might be inches. The **magnitude** is 6 inches.

Mass
The measure of the total amount of material in something.

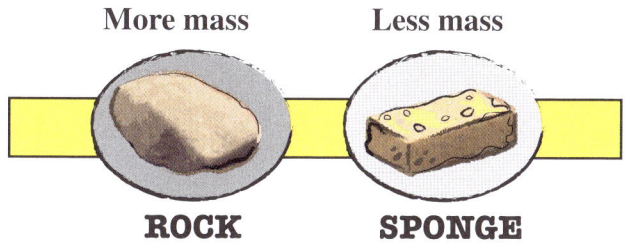

# General Science Terms

Matter
Anything that has mass and takes up space.

Anything that you can see and feel is made of **matter**.

Melt
To change from a solid to a liquid.

An ice cube will **melt** in the sun.

Model
An object built to show the details of another object that is too large or too small to see.

A **model** of the Earth is much smaller than the actual Earth. A **model** of an atom is much larger than an actual atom.

Molecule
The smallest part of a substance. A **molecule** is made up of atoms held together by covalent bonds.

Water is made of **molecules**. One **molecule** of water is the smallest amount of water you can have. It is made of two hydrogen atoms and one oxygen atom held together by covalent bonds.

General Science Terms

Nonrenewable resource
Any part of the environment that we use and cannot replace.

Fossil fuels such as coal and oil are **nonrenewable resources**. We cannot make more of them once they have been used.

Observe
To watch carefully and pay close attention to details.

Ray and Kim **observe** ducks for their science report about birds.

Particle
A part of an atom. This term can also mean a small piece of matter.

Pollute
To make the air, water, or soil unclean with waste.

Some factories **pollute** the air with smoke and chemicals.

Pollution
The waste that makes the air, water, or soil unclean.

Smoke is **pollution** that makes the air unclean.

Predict
To guess what might happen.

After watching the Moon each night for a month, we were able to **predict** when the next full moon would occur.

General Science Terms

Renewable resource
Any part of the environment that we use and can replace.

Trees are a **renewable resource**. If we plant new trees where we have taken mature ones, we will always have more trees.

Resource
A supply of something that can be used.

Resources include things, such as trees and natural gas.

Solid
Any substance with both definite shape and definite volume.

This rock is a **solid**.

Speed
A measure of the rate at which something moves.
Speed = distance ÷ time.

The **speed** of a car is 60 miles per hour. This means that for every hour of time, the car will travel a distance of 60 miles.

Substance
Any piece of matter.

Gold is a **substance**. Water is also a **substance**.

8

General Science Terms

Technology
The tools and machines that are made as a result of scientific knowledge.

When scientists discover something new, that knowledge can be used to make better machines and tools. A tool made in this way is new **technology**.

Telescope
A tool that uses lenses to help you see things that are very far away, such as other planets or stars.

Temperature
A measure of how much heat an object or area has.

Though **temperature** is always measured in degrees, scientists sometimes use different scales while measuring **temperature**. Compare the three most common **temperature** scales in the table below:

Scale	Abbreviation	Water freezes at	Water boils at
Celsius	C	0°	100°
Kelvin	K	273	373
Fahrenheit	F	32°	212°

Theory
An explanation of something that most scientists agree with.

# General Science Terms

Thermometer
A tool used to show how much heat is in a given area.

We can use a **thermometer** to measure temperature.

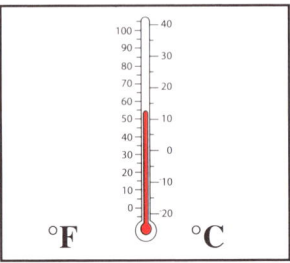

Time
A measure of how long it takes something to occur.

It takes 5 hours of **time** to fly from New York to Los Angeles.

Variable
The part of an experiment that a scientist changes in order to learn something.

In an experiment testing how well plants grow in bright sunlight, scientists might have the same plant in no light, some light, and lots of light. The **variable** in the experiment is *how much light each plant gets*.

Velocity
The speed and direction of an object.

A car's speed is 60 miles per hour, and its direction is west. The car's **velocity** is 60 miles per hour going west.

Viscosity
The measure of a liquid's resistance to flow. The higher the **viscosity**, the slower the liquid will flow.

Honey has a higher **viscosity** than water.

General Science Terms

Volume
The amount of space a three-dimensional figure takes up.

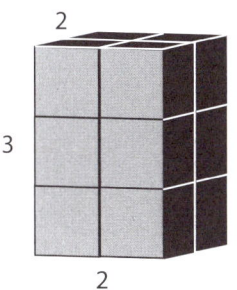

volume = 12 units

Weight
A measure of the force of gravity. An object's **weight** is a measure of how strongly the object is being pulled toward the surface of the planet.

Because **weight** is measured as a force of gravity, a person's **weight** will change depending on where they are. The Moon has less gravity than the Earth does. You will weigh less on the Moon than you do on Earth.

Biology

Abiotic
Anything that is not alive or has never been alive.

Rocks are **abiotic**.

Active transport
The movement of particles into or out of a cell that requires energy.

During **active transport**, particles move from an area of less concentration to an area of more concentration.

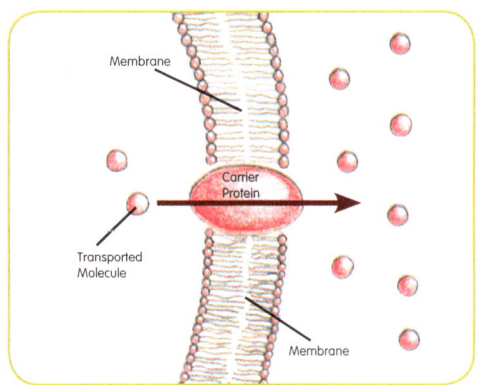

Adaptation
A trait that helps a plant or animal survive and reproduce.

Long necks are an **adaptation** of giraffes that help them eat leaves from the tops of trees.

Allele
One form of a gene that shows itself as a particular trait.

The gene for hair color has many **alleles**; one **allele** is for brown hair; another **allele** is for blonde hair.

Biology

Anaphase

The stage of mitosis and meiosis in which the chromosomes move to opposite ends of the spindle.

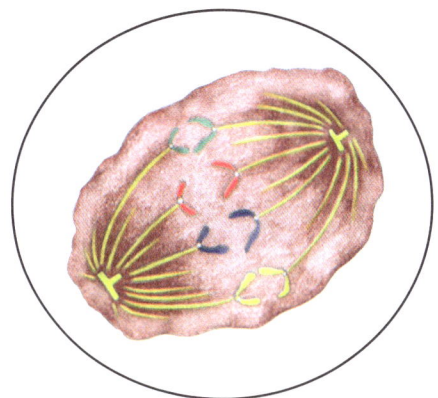

Animal

Any living thing with these characteristics:
- has cells without cell walls
- lacks chlorophyll
- cannot produce its own food
- can move around on its own

All **animals** are classified in the kingdom *Animalia*.

Antibiotic

A substance that kills bacteria.

Nicholas took an **antibiotic** to help cure his infection.

Archaea

The domain that includes a few of the most primitive one-celled organisms on Earth.

The domain **Archaea** contains only one kingdom, *Archaebacteria*. Organisms from the domain **Archaea** are most often found under extreme conditions, such as at the polar ice caps or near chemical heat vents deep in the ocean.

 # Biology

Asexual reproduction
Reproduction that requires only one parent. The offspring of **asexual reproduction** have the same genes as their parent.

Mitosis in cells is a form of **asexual reproduction**. When a cell divides, the two cells that are created have the same genes as their parent cell.

ATP
Adenosine **trip**hosphate.

ATP is the main source of energy for cells.

Bacteria (singular: bacterium)
The domain that includes most one-celled living things.

The domain **Bacteria** contains only one kingdom, called *Eubacteria*. Most **bacteria** are either helpful or have no effect on people. Some can make you sick.

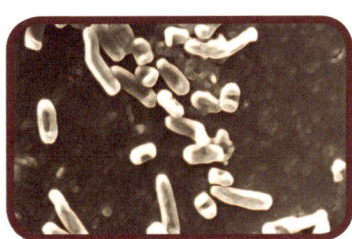

Behavior
The actions or reactions of a person or animal. The term **behavior** sometimes means *a response to a stimulus*.

Some birds, such as Canada geese, migrate south for the winter. This migration is a **behavior** of Canada geese.

Biology

Biodiversity
The number of different kinds of living things in an area.

Scientists measure how healthy an environment is by studying its **biodiversity**. An environment with a greater **biodiversity** is usually healthier than a similar environment with lower **biodiversity**.

Biogenesis
The idea that living things can only come from other living things, not from nonliving matter.

The theory of **biogenesis** states that a living organism can only be made from another living organism. Life cannot come from things that are not alive, such as rocks or water.

Biological transport
The movement of substances, such as nutrients, into or out of a cell. **Biological transport** can also mean the movement of substances from one cell to another.

The blood cells in your body move nutrients from your digestive system to other organ systems. This is an example of **biological transport**.

Biosphere
All the parts of the Earth where things can live.

15

Biology

Biotic
Anything that is alive or has ever been alive.

Camouflage
A color or pattern on an animal that lets it blend into its environment.

Cancer cells
Cells that begin to reproduce very quickly and invade other areas of the body.

Cancer cells can develop into tumors that invade other areas of the body, including the bloodstream.

Carbohydrate
A compound that is an energy source for living things. **Carbohydrates** are made of carbon, hydrogen, and oxygen.

Starch is an example of a **carbohydrate**. It can be found in common foods like potatoes and bread.

Carbon
An element that is found in many abiotic and in all biotic compounds.

Biology

Carbon cycle

The path that carbon takes as it moves throughout the biosphere.

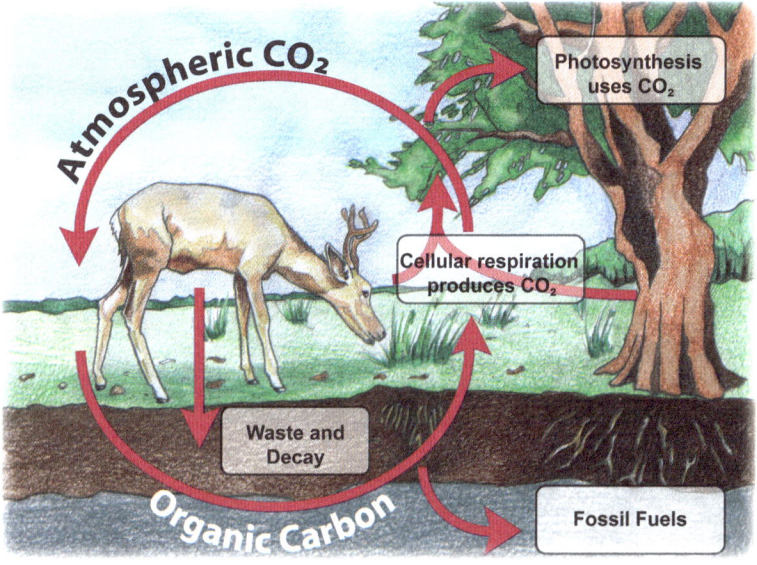

Catalyst

A substance that helps two or more other substances react and change. The **catalyst** does not change when it helps the other substances react.

Catalyst

# Biology

Cell
The basic unit of life. All living things are made of one or more **cells**.

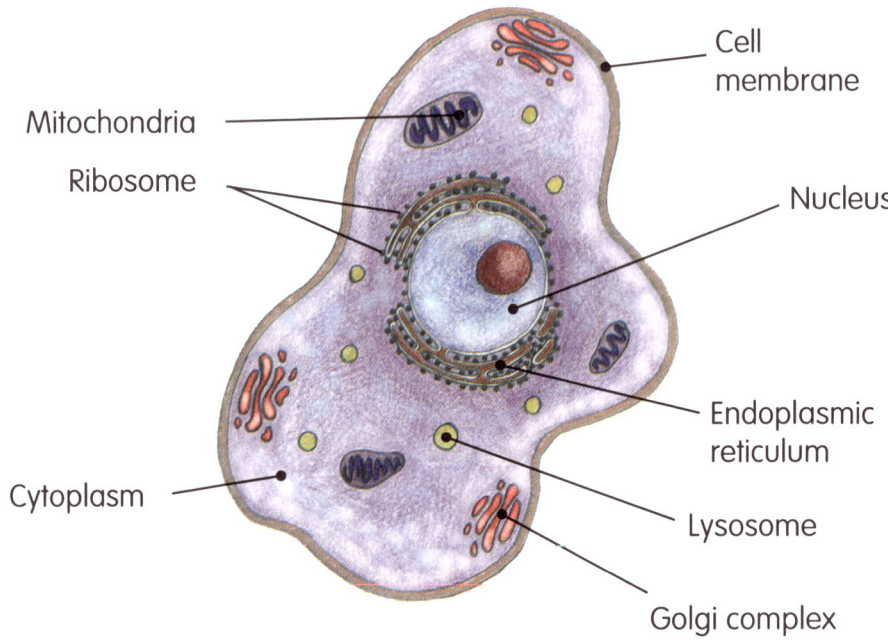

Cell differentiation
The process by which a cell becomes specialized in order to do a specific job, as in the case of a liver cell or a blood cell.

There are many different cells in the human body, such as heart cells, kidney cells, brain cells, and skin cells. All cells in one human's body contain exactly the same DNA.

Biology

Cell membrane

The outside layer of a cell. The **cell membrane** allows nutrients into the cell and lets waste out. It separates the cell from the environment.

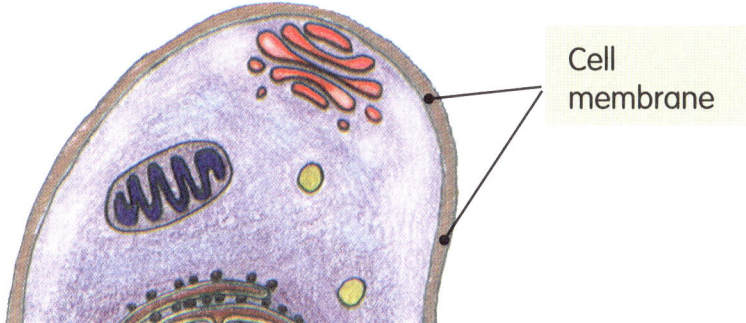

Cell wall

The boundary between plant cells and the outside world.

The **cell wall** is harder than a cell membrane, so it gives the plant cell structure.

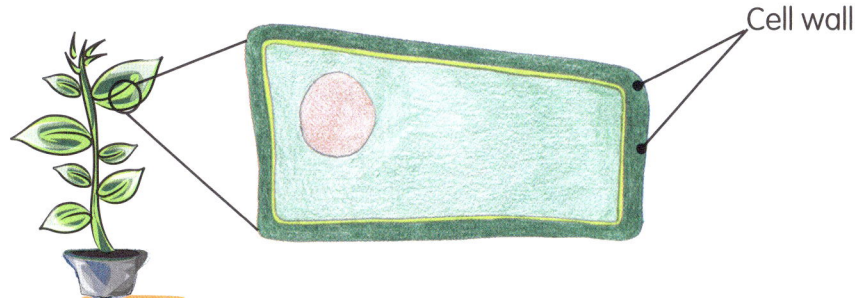

Chlorophyll

The green substance that gives most plants their color and allows them to turn water, sunlight, and carbon dioxide into food.

Chlorophyll provides energy for all green plants.

Biology

Chloroplast
Plant cell organelle that holds the chlorophyll that makes food for the plant.

Only plant cells have **chloroplasts**, because these organelles are necessary for turning sunlight, water, and nutrients into energy.

Chromosome
A strand of DNA that carries genes.

Humans have 23 pairs of **chromosomes** in each cell of their bodies.

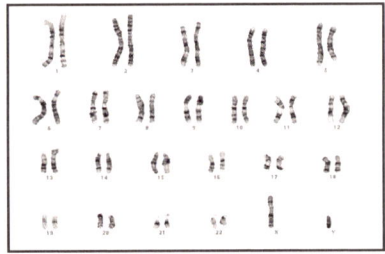

Cilia (singular: cilium)
Tiny hair-like parts that extend out from the membranes of some one-celled organisms. The **cilia** wave back and forth and help the organism to move.

The **cilia** on a paramecium help it to move through the water.

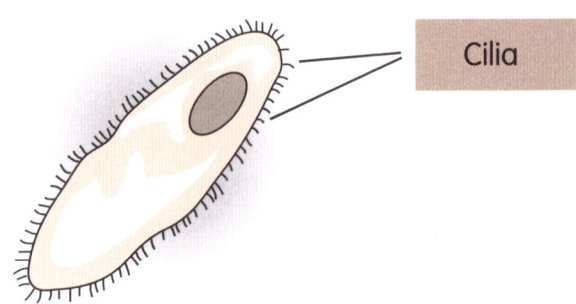

Biology

Citric acid cycle
A series of reactions that turn food into energy for the cell. This cycle is also called the *Krebs cycle*.

The **citric acid cycle** occurs in the mitochondria of cells.

Classification
The grouping of things that are alike.

The **classification** of the plants shown here are into a group called *flowers*.

Community
All the populations of living things that share an area.

A forest **community** can include trees, grass, birds, animals, and many other living things.

Consumer
A living thing that eats plants or animals.

All animals are **consumers**.

Biology

Cytokinesis
The division of the cytoplasm of a cell after the division of the nucleus.

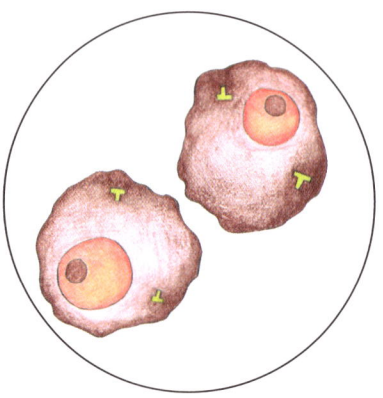

Cytoplasm
Material that is inside a cell. The **cytoplasm** does not include the nucleus.

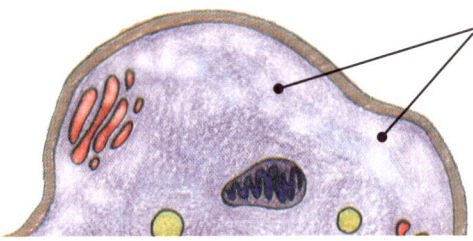

Daughter cells
The offspring cells that are the result of mitosis.

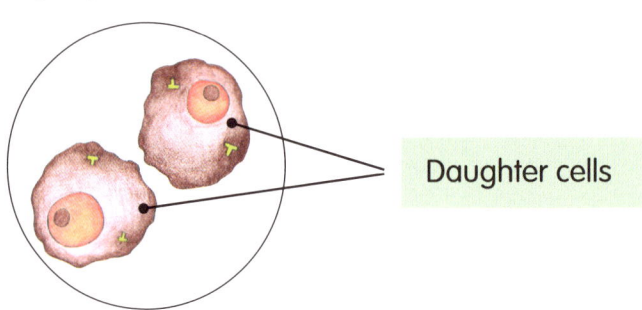

Biology

Dead organic material
What is left of something that was once alive.

Decomposer
A living thing that breaks down animal waste, dead animals, and dead plants for food.

This vulture is an example of a **decomposer**.

Diffusion
The movement of particles from an area where there are many of that kind of particle to an area where there are not many of that kind of particle.

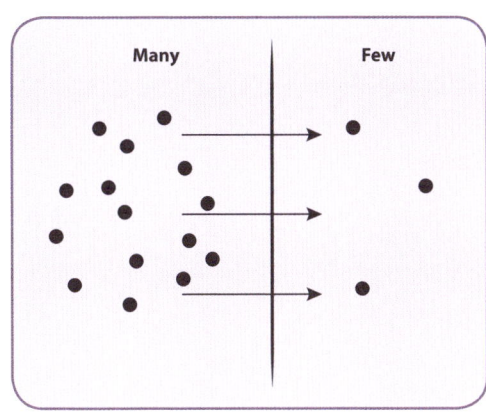

23

Biology

Diversity
Having many different types of organisms in an area.

Rainforests have a lot of **diversity**. This means rainforests contain many different species of living things.

DNA
A molecule found in all living things. This molecule passes information a living thing needs to grow and function from parents to offspring. **DNA** is grouped into structures called chromosomes.

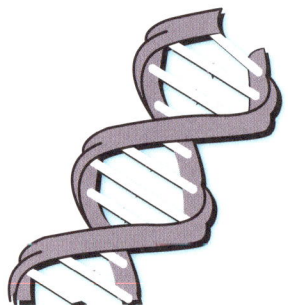

DNA stands for *deoxyribonucleic acid*.

Domain
The top-level grouping of living organisms.

Three Domains
1. Bacteria
    contains bacteria
2. Archaea
    contains archaeobacteria
3. Eukarya
    contains animals, plants, fungi, and protists

Biology

Dominant
Term used to describe an allele that always expresses when it is present.

Ecology
The study of the relationships between organisms and their environments.

A scientist who studies **ecology** is called an *ecologist*.

Ecosystem
All the living and nonliving things in an environment. A pond **ecosystem** can include plants, frogs, water, rocks and insects.

Electron transport chain
The movement of electrons from one reaction to another as nutrients are changed into ATP.

Embryo
The first stages of development of a multi-cellular organism.

In most mammals, the **embryo** grows inside the mother organism. In birds, the **embryo** develops in an egg.

25

Biology

Endangered species
A species that is in danger of becoming extinct.

The Red Wolf is an **endangered species**. There are fewer than 300 left in the world.

Endoplasmic reticulum
Membranes found in a cell's cytoplasm. The **endoplasmic reticulum** help the cell create and transport proteins and lipids.

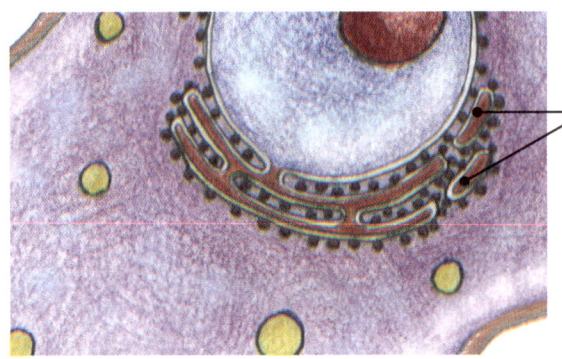

Endoplasmic reticulum

Environment
Everything surrounding a living thing.

An **environment** includes living and nonliving things as well as conditions such as temperature, precipitation, and climate.

Environmental influence
The effect of the environment on a living thing.

Because of **environmental influence**, some mammals in the Arctic Circle have thick fur that keeps them warm.

Biology

Enzyme

A type of protein that helps other organic substances change and react. The **enzyme** itself does not change. **Enzymes** are a type of catalyst.

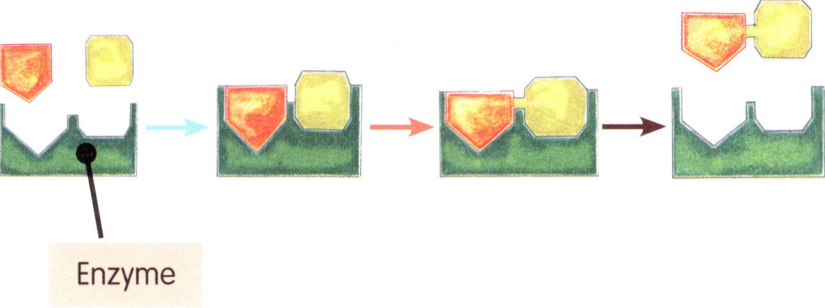

Enzyme

Eukarya

The domain that includes all higher-functioning organisms.

There are four kingdoms in **Eukarya**: *Animals*, *Fungi*, *Plants*, and *Protists*. All organisms from **Eukarya** have *eukaryotic* cells, which are cells that contain a distinct membrane-bound nucleus.

Evolution

The changes a species goes through over time.

Small prey, such as mice, have evolved to have coloring that is similar to their surroundings. This **evolution** has helped them to survive.

External stimulus

Something outside an organism that makes it respond.

For a predator such as a snake, an **external stimulus** might be a hopping frog. The **external stimulus** may get the snake to hunt the frog.

27

Biology

Extinct
No longer existing or living.

Dinosaurs became **extinct** long ago.

Fermentation
A breakdown of nutrients to make energy that does not use oxygen.

Yeasts are organisms that use **fermentation** to generate energy. Yeasts give off alcohol as a result of **fermentation**, which people use to make wine.

Fertile
Able to reproduce; also, able to help other things grow and develop.

Frogs become **fertile** when they reach the adult stage. **Fertile** soil has nutrients that help plants grow.

Fertilization
The process where a male gamete (sperm) combines with a female gamete (egg) during sexual reproduction.

Flagellum (plural: flagella)
A long, tail-like extension at the end of a one-celled organism that helps the organism move.

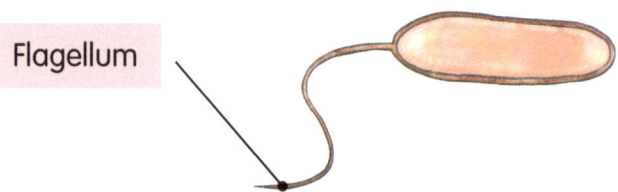

Biology

Food chain
The food and energy relationships among living things.

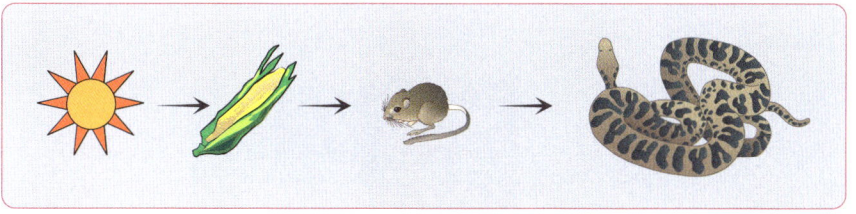

Food web
The connection of many different food chains in an ecosystem.

Foreign substances
Something that is not naturally found in an organism.

Foreign substances like poisons can make an organism very ill.

Fossil
Evidence of a living thing that died thousands or millions of years ago.

Fossils include bones, shells, and impressions such as footprints.

Fossil record
A term used to refer to all the fossils that have been discovered. This term also refers to the information we have learned from all the fossils.

29

Biology

Fungi (singular: fungus)
The kingdom that includes mushrooms, molds, and yeasts. Fungi are **decomposers**.

Gamete
A mature sperm or egg cell.

Gametes are sometimes called *sex cells*.

Gene
A small part of a DNA molecule that carries the hereditary information for one trait.

You have a **gene** that tells what your eye color is.

Gene expression
The process by which a gene tells a cell what to do or how to grow. Each cell expresses only a small part of its genes.

A muscle cell will express different genes than a liver cell.

Gene mutation
Any change to the sequence of bases in DNA.

Sometimes, **gene mutations** can change the way an organism looks or acts.

Biology

Gene pool
The genetic information in a whole population.

The **gene pool** is one way to describe all the different types of genes each member of a species has.

Genetic drift
Random changes in how often a trait appears in a small isolated population.

Genetic engineering
Changing the way an organism functions by making changes to the organism's DNA.

Some scientists think that the **genetic engineering** of corn will make it easier to grow.

Genetics
The study of gene variation and inheritance.

A scientist who studies **genetics** is called a *geneticist*.

Genus
The classification of living things that is just above species. Organisms from the same **genus** have similar characteristics, but they cannot mate and produce fertile offspring.

The **genus** *Felis* contains many species, including *Felis concolor*, the cougar, and *Felis catus*, the house cat.

Biology

Glucose
The most basic sugar.

Glucose is the major energy source for the body.

Glycolysis
A process in which glucose is changed into pyruvic acid. ATP is created as a result of this process.

Glycolysis is one way of producing energy for an organism.

Golgi complex
A cell organelle that helps prepare material to be moved out of the cell.

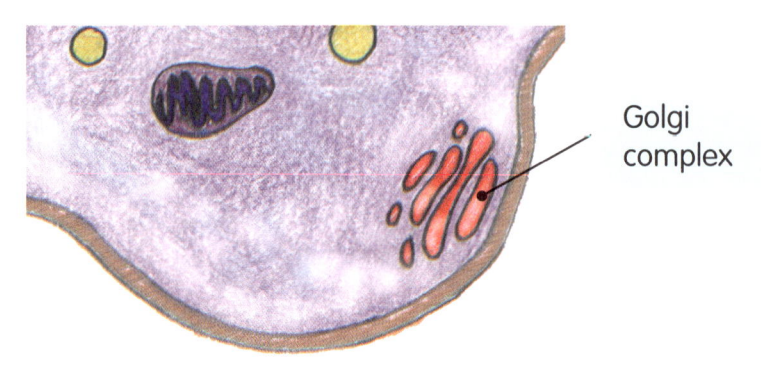

Grow
To get bigger or more complex.

Biology

Habitat
The place where a living thing is commonly found.

This frog's **habitat** is a tree.

Heredity
The biological similarity of parents and offspring.

Most people can find features of their own body that look like their parents. This is because of **heredity**.

Hormone
A substance that is made by cells in one part of the body and that affects the behavior of cells in a different part of the body.

The **hormone** *melatonin* is produced in the brain. It travels throughout the body and is responsible for controlling the body's sleep cycle.

Human being
A member of the human race.

All people are **human beings**.

Biology

Immune system

The system that keeps the body safe from disease and foreign substances.

Skin is an important part of the body's **immune system**. It keeps many foreign substances and diseases from entering the body.

Independent assortment

The random sorting of chromosomes during sexual reproduction that produces many different combinations of genes in the offspring.

Independent assortment is how different sex cells from the same person can have different combinations of genes.

Inhabit

To live somewhere.

This deer **inhabits** a forest.

Inherited characteristics

Traits such as eye color that are passed from parents to offspring.

Inherited characteristics are passed from parent to child in genes.

Biology

Interphase
The stage of a cell between cell divisions.

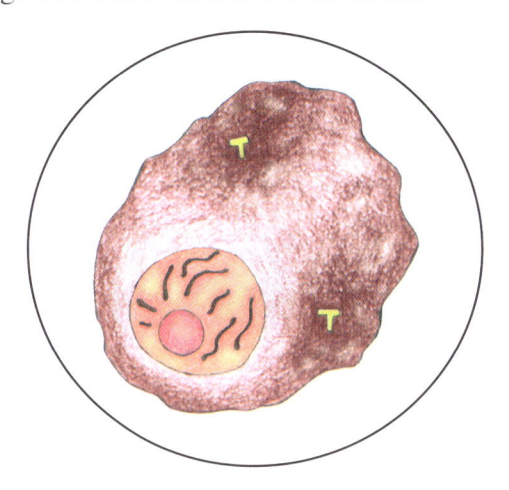

Introduced species
A species that has been brought by humans to a location where it did not occur naturally.

Honeybees are an **introduced species**. They were brought to the United States by people, and have become an important part of the U.S. ecosystem.

Invasive species
A species that has been brought by humans to a location where it did not occur naturally. This kind of species becomes a pest in the new location, threatening the local biodiversity.

The snakehead fish is an **invasive species**. It was brought to the United States by people, and has begun destroying local fish populations.

Biology

Kingdom

The second-level grouping of organisms. A group of **kingdoms** make a domain.

> Three Domains
>
> *Bacteria*
>   **Kingdom:** *bacteria*
>
> *Archaea*
>   **Kingdom:** *archaeobacteria*
>
> *Eukarya*
>   **Kingdoms:** *animals, plants, fungi, and protists*

Life cycle

The biological events an organism goes through as it reaches sexual maturity.

The **life cycle** of a frog begins with eggs and ends when those eggs become new adult frogs that lay eggs of their own.

Biology

Lipids
A group of organic compounds, including the fats, oils, and waxes, that are oily to the touch.

Lipids are one of the basic materials of living cells.

Lysosome
A cell organelle that helps the cell digest food.

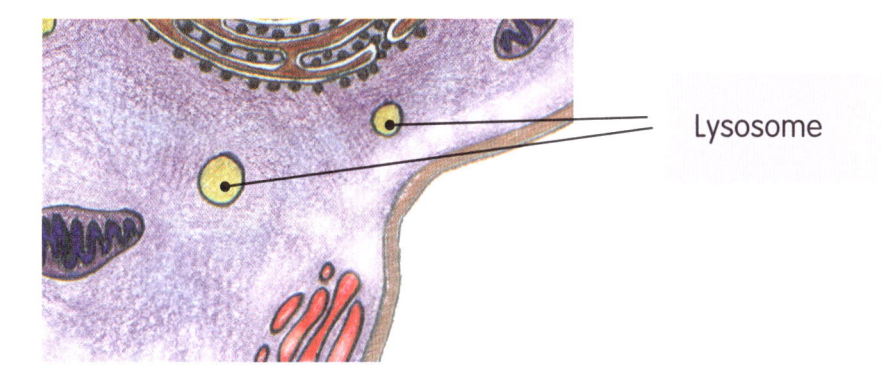

Lysosome

Macromolecule
A very large molecule, such as a polymer or protein.

Macromolecules are made of many smaller identical units linked together.

"Macro-" is a prefix that means "big."

Marine organism
An organism that lives in or around salt water.

37

Biology

Meiosis

A cell division process that creates sex cells (sperm or eggs) in sexually reproducing organisms. This process also results in the creation of mold spores in certain molds.

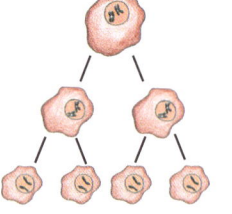

When a cell goes through **meiosis**, it passes through each phase of *mitosis* (see page 40) two times, so that the cell divides into four gametes rather than two daughter cells.

Membrane

A thin layer of tissue that covers surfaces or separates structures and organs of an animal or a plant.

Cells have an outer **membrane**. Parts of the human body, such as the inside of the mouth, are covered in **membrane** as well.

Mendel's laws of heredity

The "laws" that explain how traits are passed from parents to offspring. These laws were first discovered by Gregor Mendel, who learned about them by studying pea plants.

1. The *law of segregation* says that when a sexually reproducing organism creates sex cells through meiosis, each pair of alleles separates, so that only half of each allele pair becomes part of the new sex cell. Because of this, any parent organism that has one dominant gene and one recessive gene may pass the dominant gene to one offspring and the recessive gene to another.
2. The *law of independent assortment* says that every pair of alleles separates independently of every other pair of alleles during meiosis. Because of this, an organism can create a wide variety of possible genetic combinations in the sex cells.

Biology

Metabolism
The chemical processes that a living thing needs to do to stay alive.

In **metabolism**, some substances are broken down to create energy. Other substances that are necessary for life are also created.

Metamorphosis
A change in the form and habits of an animal as it grows from birth to an adult.

Some examples of **metamorphosis** include the transformation of a caterpillar into a butterfly and the changing of a tadpole into a frog.

Metaphase
The stage of mitosis and meiosis during which the chromosomes are aligned along the **metaphase** plate.

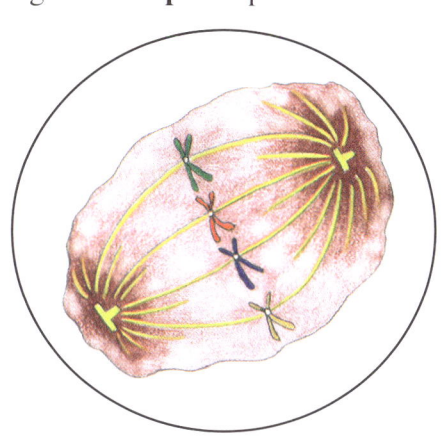

Biology

Microorganism
An organism that is too small to see with your eyes alone. These are sometimes called *microbes*.

Microorganisms include bacteria, some algae, and yeasts.

Migration
The movement of groups of animals (especially birds or fish) from one region to another for feeding or breeding.

Canada geese follow a **migration** path south for the winter.

Mitochondria
Organelle that gives energy to the cell.

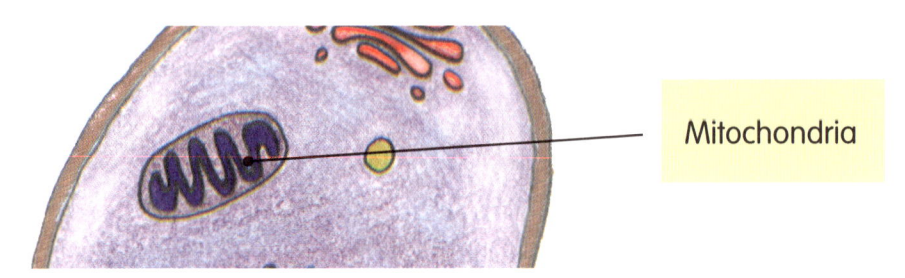

Mitosis
The process of nuclear cell division. There are typically five stages of **mitosis**: interphase, prophase, metaphase, anaphase, and telophase.

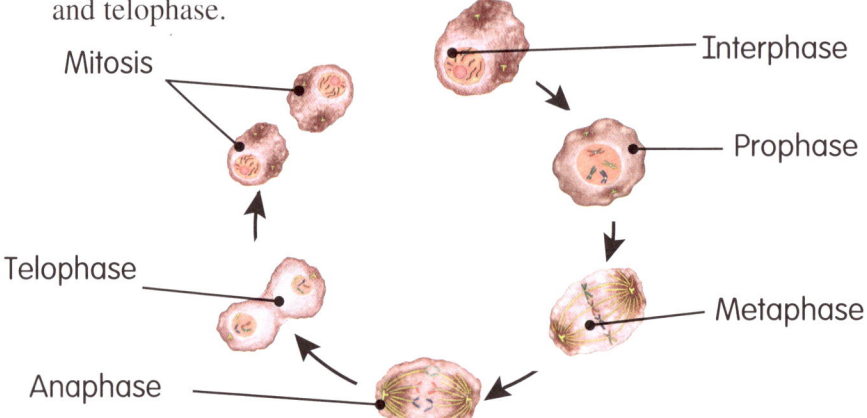

40

Biology

Native species
A species that occurs naturally in a certain environment.

Giant pandas can be found in zoos all over the world, but they are a **native species** of China.

Natural selection
The theory that says if a living thing has traits which help it survive in its environment, it is likely to live longer and have more offspring. It passes those helpful traits to its offspring and they pass it to their offspring. Eventually, the helpful traits become more common.

Through **natural selection**, giraffes have developed long necks that help them eat leaves from tall trees.

Nonliving
Never having been alive. Also see abiotic.

Water is **nonliving**. Rocks are also **nonliving**.

Nucleic acids
A large molecule that is found in living things and is made of groups of smaller molecules called nucleotides.

DNA is one type of **nucleic acid**.

Nucleotide
The basic molecule that forms DNA and RNA.

Biology

Nucleus

The part of the cell that holds the cell's DNA.

The plural of **nucleus** is *nuclei*.

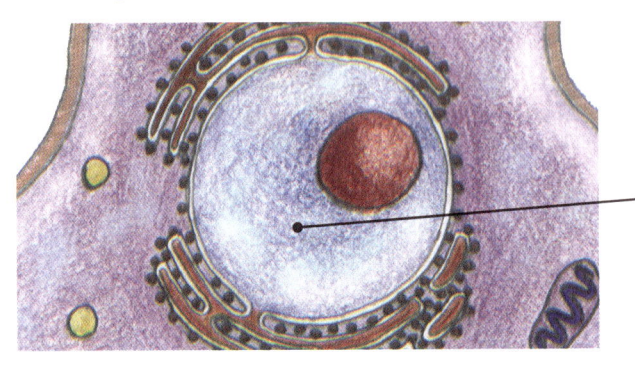

Nucleus

Nutrient

A substance that the body needs in order to live.

Nutrients include carbohydrates, fats, proteins, vitamins, and minerals.

Offspring

The "child" of a living thing. **Offspring** get their traits from their parents.

Offspring

Biology

Organ
A part of the body that performs a certain job and is made of two or more kinds of tissues.

The heart is an **organ**. It is made of muscle and other tissues. It performs the job of moving blood through the body.

Organ system
A group of organs that work together.

The heart is part of an **organ system** called the *cardiovascular system*. This system also includes arteries and veins.

Organelle
A cell part that performs a certain job.

One of the **organelles** reviewed in this book is the *mitochondria*, which is the **organelle** responsible for generating energy in a cell.

Organism
An individual form of life, such as a plant, animal, bacterium, protist, or fungus.

An **organism** can be made of a single cell or many cells.

Parent
An organism that produces offspring.

Parent

Biology

Passive transport

Movement of a substance across a cell membrane without using any of the cell's energy, as in diffusion.

Water often enters cells through **passive transport**.

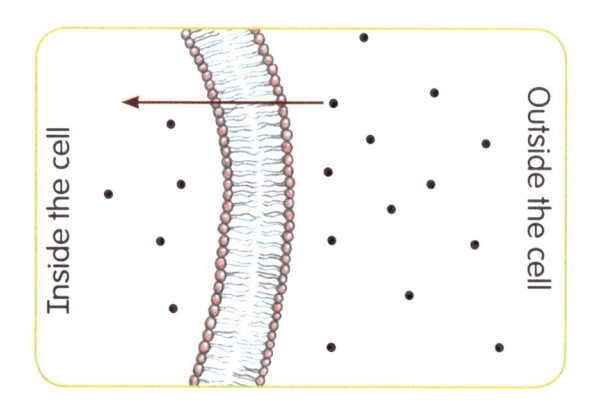

Pathogen

Anything that causes disease, especially a living microorganism such as a bacterium or fungus.

Pathogens can enter the body through a cut on the skin, or by eating tainted food.

Photosynthesis

The process that green plants use to turn sunlight, water, and carbon dioxide into food.

"Photo-" is a prefix that means "light."

Biology

Plant
The kingdom of organisms that make their own food from sunlight, have cell walls, and are not usually able to move around.

Plants include grass, moss, ferns, and trees.

Pollen
A small particle that is produced by a seed plant and contains the male sex cell of the plant.

Because they cannot move around, plants rely on wind and animals such as bees to transport their **pollen**.

Pollination
The transfer of pollen from the male part of a seed plant to the female part of a seed plant.

Bees help with **pollination** by carrying *pollen* from one flower to another.

Animals that assist in **pollination** are known as *pollinators*.

Population
A group of the same species of plant or animal living in the same place.

A **population** of bullfrogs will include each individual bullfrog that lives in one area, such as near a pond.

Biology

Predator
An organism that hunts and eats other organisms.

Owls are **predators**. They hunt animals such as mice for food.

Predator

Prey
An organism that is hunted and eaten by another organism.

Mice are **prey**. They can be hunted and eaten by other animals, such as owls.

Prey

Producer
An organism that can make its own food.

Plants are **producers**. They make their own food by the process of photosynthesis.

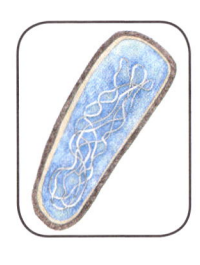

Prokaryote
A type of single-celled organism.

Prokaryotes do not have a distinct, membrane-bound nucleus or membrane-bound organelles. Their DNA is not organized into chromosomes.

Biology

Prophase
The first stage of mitosis during which the chromosomes condense and become visible, the nuclear membrane breaks down, and the spindle forms at opposite ends of the cell.

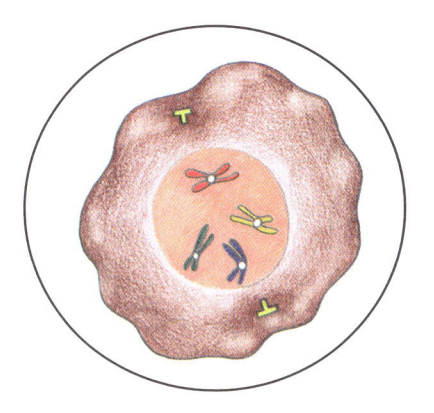

Protein
Complex organic macromolecules that contain carbon, hydrogen, oxygen, nitrogen, and sometimes sulfur.

Proteins are made of one or more chains of amino acids.

Protist
An organism from the kingdom *Protista*.

Protists can be one-celled or multicellular. Some make their own food, while others feed on other organisms. All **protists** lack complex organ systems.

Recessive
Term used to describe an allele that only expresses itself when there is no dominant allele in the cell.

Recycle
To extract useful materials from garbage or waste.

Using food scraps as compost in a garden is one way for many people to **recycle** their waste.

Biology

Reproduction
The process that allows one species to make more of itself.

Reproduction can be asexual or sexual.

Respiration
The process by which an organism takes in and lets out gases such as oxygen and carbon dioxide.

Respiration is sometimes called *breathing*.

Ribosome
Cell part that makes protein.

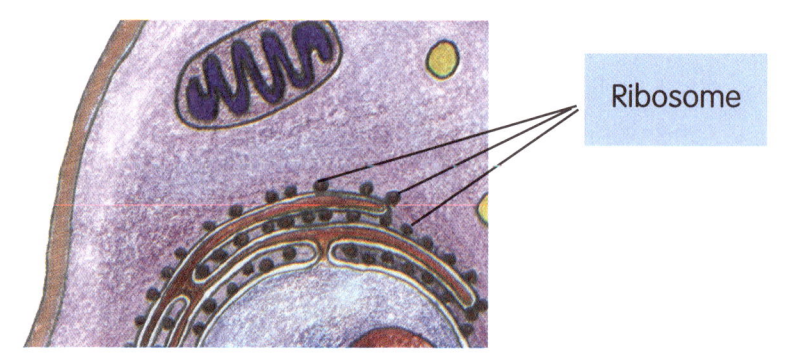

Ribosome

RNA
A single strand of compounds that takes information from the DNA and brings the information to other parts of the cell.

RNA stands for *ribonucleic acid*.

Segregation
The separation of a pair of alleles so that each one appears in a different sex cell.

Segregation is when a pair of alleles separates.

Biology

Sexual reproduction
Reproduction that requires two parents.

The offspring of **sexual reproduction** receive half of their genes from one parent and half from the other parent.

Species
A group of organisms that can produce fertile offspring.

The tiger is one **species** of big cat.
There are many other **species** of big cats.

The word **species** is both a singular and a plural word.

Spindle
A system of tiny fibers that help the chromosomes separate during cell division.

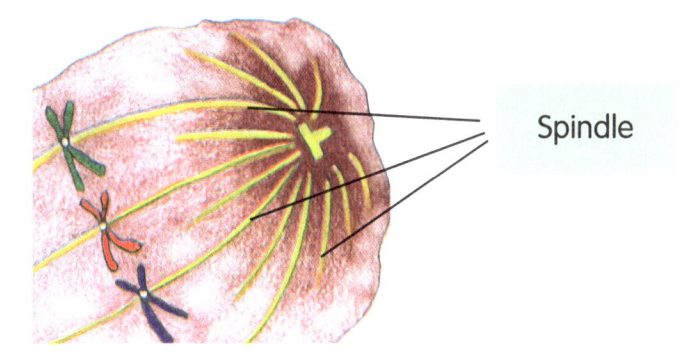

Spindle

Stimulus (plural: stimuli)
Something that provokes an organism to respond.

A **stimulus** can be external, such as when a mouse causes a cat to begin chasing it. A **stimulus** can also be internal, such as when the feeling of being tired causes you to go to bed.

49

Biology

Taxonomy
The study or development of ordered classification systems.

A scientist who studies **taxonomy** or who develops a new way of classifying things is called a *taxonomist*.

Telophase
The final stage of mitosis or meiosis during which the chromosomes of daughter cells are grouped into new nuclei.

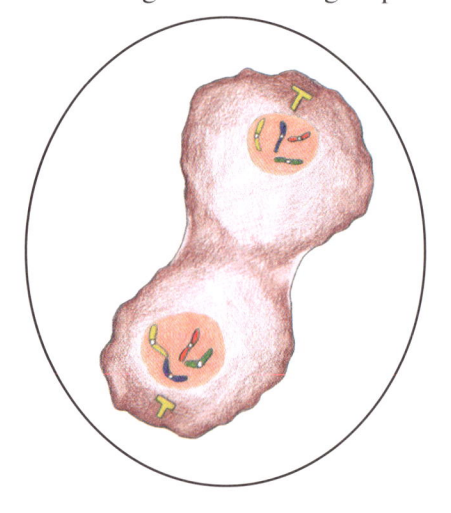

Tissue
A group of similar cells that work together to do a similar job in the body.

There are four basic types of **tissue**: muscle, nerve, epidermal, and connective.

Trait
A characteristic that is inherited.

Characteristics such as eye color and height are **traits** that get passed from parent to offspring.

Biology

Uncontrolled cell division
Cell division that is not normal for a healthy body.

Uncontrolled cell division can eventually form a tumor. This tumor may be *malignant* (cancerous) or *benign* (not cancerous).

Vaccine
A medicine made of a weakened or killed bacterium or virus. This medicine does not contain enough of the bacterium or virus to make a person sick, but it trains a person's body to be able to recognize and fight off a later infection of that bacteria or virus.

Many people receive **vaccines** against diseases when they are small children.

Vacuole
Cell part that stores waste or food.

Vacuoles are most often found in plant cells.

Virus
Simple microscopic parasites that are made of a core of RNA or DNA surrounded by a protein coat. Since **viruses** cannot reproduce on their own, most scientists do not consider them to be alive.

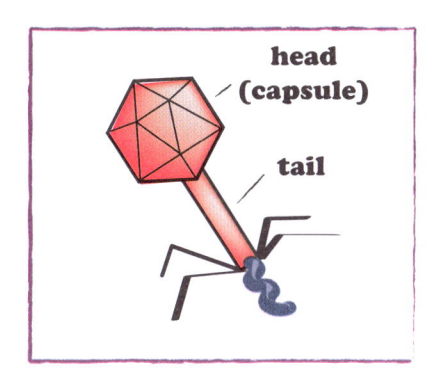

Biology

Waste materials

Anything that is left over after a living thing has used something.

Food waste, fecal matter, and urine are examples of **waste materials**.

Zygote

A cell that results from the fusion of a sperm cell and an egg cell.

A **zygote** is a fertilized egg.

Chemistry

Acid
A solution that contains hydrogen ions (H⁺).

Lemon juice is an **acid**. Anything that is an **acid** can be called *acidic*.

Atmospheric pressure
A measure of the force exerted by the atmosphere on living and nonliving things.

Atmospheric pressure is about 14.7 pounds per square inch (14.7 psi) at sea level.

Atom
The smallest unit of an element. **Atoms** can be broken down into smaller particles, but when this happens, they stop having the properties of a particular element.

All **atoms** are typically made up of protons, neutrons, and electrons.

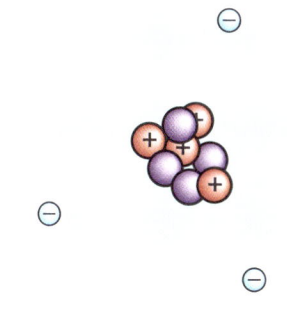

Atomic mass unit (amu)
A unit of measurement that tells the mass of atoms.

An atom's exact **AMU** can be found by looking at a periodic table of the elements.

Chemistry

Atomic number
The number of protons in the nucleus of each atom of an element.

Oxygen atoms have 8 protons. Oxygen's **atomic number** is 8.

Avogadro's law
If you have equal volumes of two gases, and they are at the same temperature and pressure, then the two gases contain the same number of molecules.

The two balloons below are the same size, same temperature, and are under the same pressure. One contains oxygen. The other contains helium. But both contain exactly the same number of molecules. This is an example of **Avogadro's law**.

Avogadro's number
The number of particles in exactly one mole of a substance.

Avogadro's number equals 6.02×10^{23}.

Barometer
A tool used to measure atmospheric pressure.

A **barometer** is an important tool for predicting the weather.

Chemistry

Base
A solution that contains hydroxide ions (OH-).

Bleach is a **base**. Bases are sometimes called *alkali*. Anything that is a **base** can be called *alkaline* or *basic*.

Boiling point
The temperature at which a liquid quickly changes into a gas.

The **boiling point** of water is 212° F or 100° C.

Carbon
An element consisting of six protons, six neutrons, and six electrons.

Carbon is found in many different forms, including diamonds and coal, and in many compounds, including methane and carbon dioxide.

Catalyst
A substance that changes the rate of a chemical reaction. **Catalysts** are not changed when they act on other substances.

Catalyst

Chemistry

Charged particle
A particle that has a positive or a negative charge.

Protons are positively **charged particles**. Electrons are negatively **charged particles**.

Chemical
An element or a compound.

All **chemicals** have a definite composition, which means they are always made of the same amounts and kinds of atoms.

Chemical bond
The force that holds two atoms together.

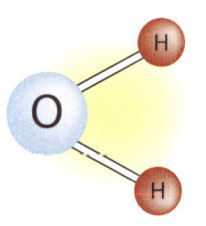

In water, hydrogen forms **chemical bonds** with oxygen.

Chemical change
A change that a substance undergoes when it becomes a new substance.

In a **chemical change**, chemical bonds between atoms break and new substances are formed.

Chemical formula
A system of describing chemicals. Each chemical is shown as a formula based on the types and number of atoms that make up the chemical.

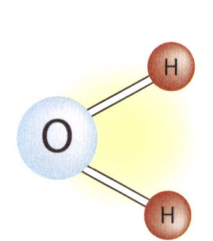

Water is described with the **chemical formula** H_2O because each molecule contains two hydrogen atoms and one oxygen atom.

Chemistry

Chemical potential energy
The energy contained in the chemical bonds of a substance.

The energy in chemical bonds is called *potential* because it is not being used. It is stored, like a rubber band stretched tight. When the energy is used by breaking the bonds, it is like letting the rubber band fly across the room.

Chemical property
The ability of a substance to react in the presence of another substance.

Iron has the **chemical property** of being able to react with oxygen. We can see this reaction as iron turns to rust.

Chemical reaction
The process of one or more substances changing into other substances.

Iron changes to rust when exposed to oxygen. The **chemical reaction** of iron and oxygen results in rust.

Conductivity
The ability of an element or compound to carry electricity or heat.

A metal pot on the stove becomes hot very quickly. It has a high **conductivity**, and it can be described as *conductive*. Wood is not as *conductive* as metal.

Chemistry

Conservation of mass

The principle stating that mass cannot be created or destroyed.

The principle of **conservation of mass** says that the mass of the reactants in an experiment should always equal the mass of the products. The principle is a useful tool for checking your work in a chemical equation.

Covalent bond

The bond created by two atoms when they share one, two, or three pairs of electrons.

Water is made of oxygen and hydrogen atoms held together by **covalent bonds**. In many diagrams, **covalent bonds** are represented by a line or stick shape.

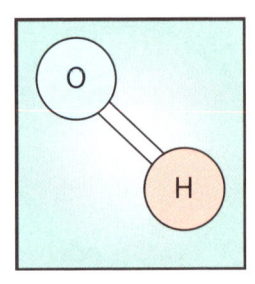

Crystal

A substance in which the particles are in an ordered, repeating geometric pattern.

Substances that have the properties of a **crystal** are called *crystalline*.

Chemistry

Electron
A particle of an atom that carries a negative charge and moves in a cloud around the nucleus.

Electron transfer reaction
Please see *oxidation-reduction reaction*.

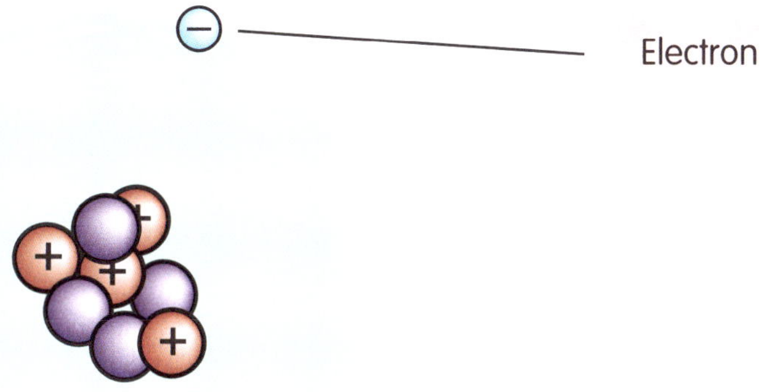

Electron

Element
A substance made of atoms that are all the same.

Every **element** that has been discovered is listed in the periodic table of the elements. (Please see the Periodic Table on page 128 of the Reference section.)

Endothermic
Any process in which heat is absorbed.

The melting of ice into water is an example of an **endothermic** process. Frozen water, or ice, absorbs heat and becomes liquid water.

Chemistry

Equilibrium
The state in which varying, shifting, or opposing forces are in balance.

The prefix "equi-" means "equal."

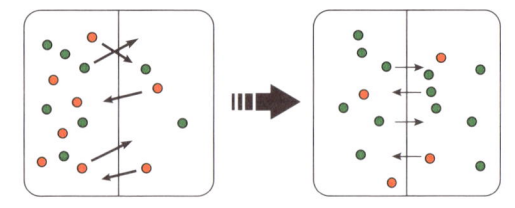

Exothermic
Any process in which heat is released.

The freezing of water into ice is an example of an **exothermic** process. Liquid water releases heat, and becomes solid water, or ice.

Freezing point
The temperature at which a liquid turns into a solid.

The **freezing point** of water is 32° F and 0° C.

Functional groups
An atom or group of atoms that determines the properties of an organic compound.

Each of the molecules shown below contains an O-H **functional group**, called the alcohol **functional group**. The O-H **functional group** causes each of the molecules to act in a similar way.

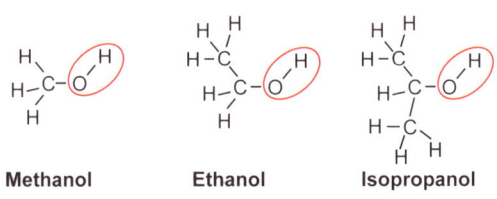

Methanol   Ethanol   Isopropanol

60

Chemistry

Gas laws
Mathematical relationships between the temperature, volume, quantity, and pressure of a gas.

Gas volume
The space taken up by a gas.

Pressure is an important factor in **gas volume**. The jar at right shows oxygen molecules under pressure. The jar at the far right shows oxygen molecules at a greater pressure. The same gas molecules take up less space due to the increase in pressure.

Gaseous
Having properties of a gas.

Helium is a **gaseous** element. It normally exists as a gas.

Heat flow
The process of heat moving from an area of higher temperature to an area of lower temperature.

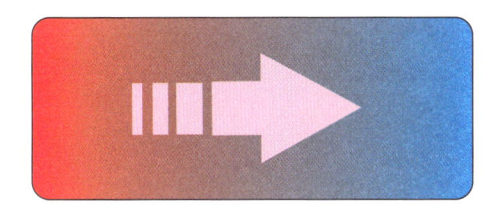

Hydrogen
The lightest element. **Hydrogen** is made of one proton and one electron.

There are more **hydrogen** atoms than any other kind of atom.

61

Chemistry

Hydrogen bond
A molecular bond that connects hydrogen atoms to other atoms such as oxygen.

Hydrogen bonds connect hydrogen to oxygen on H_2O.

Ideal gas
An imaginary gas that fits all the assumptions of the kinetic molecular theory.

Scientists use the **ideal gas** to help them think of new solutions to problems.

Inorganic
Compounds that do not contain carbon-to-hydrogen bonds.

Water is **inorganic**. It does not contain any carbon.

Ion
An atom that has gained or lost at least one electron. **Ions** have either a negative or a positive charge.

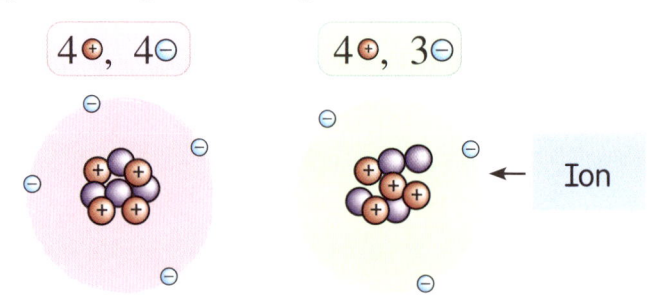

Ionic bond
A bond between ions that results from electrical attraction.

You can imagine an **ionic bond** as being like the attraction between opposite poles of two magnets.

Chemistry

Kinetic molecular theory
A theory that says molecules of a gas are always in motion. This theory also states that the molecules in a gas are usually very far apart from one another.

Remember that *kinetic* means "in motion." This may help you remember that the **kinetic molecular theory** is a theory about how molecules are in motion.

Latent heat
The heat that is absorbed or released by an element during a change of state, such as solid to liquid, or liquid to vapor.

When **latent heat** is absorbed or released, the temperature of the substance being heated does not change.

Melting point
The temperature at which a solid melts into a liquid.

Chocolate has a very low **melting point**. Most kinds of chocolate melt at around 60°–80° F. This is why chocolate can melt in your mouth.

Metal
A substance that is shiny, conducts heat and electricity, and bends without cracking.

63

Chemistry

Mixture

A combination of two or more substances. In a mixture, the substances do not bond to each other or change.

You can create a **mixture** of oil and water. If you shake up oil and water, they seem to come together, but the molecules never bond together.

Molarity

The number of moles of solute in one liter of solution.

Imagine you have 2 liters of water that contain 0.5 moles of dissolved salt. To figure out the **molarity** of the solution, you simply divide the number of moles of solute by the total amount of mixture.

0.5 moles of salt
2 liters of solution

This solution has a **molarity** of 0.25 mol.

Mole (mol)

A unit of measurement of the amount of a substance.

The **mole** is a standard unit of measurement. One **mole** is equal to the number of atoms in 12 grams of carbon-12, so each **mole** is equal to a large number of atoms. The **mole** is very useful to scientists. It would take quite a long time to count the number of individual atoms in a sample of an element. By counting in **moles**, scientists can measure much more quickly.

Chemistry

Neutron
A particle found in the nucleus of an atom. **Neutrons** have no charge and are roughly the same mass as a proton.

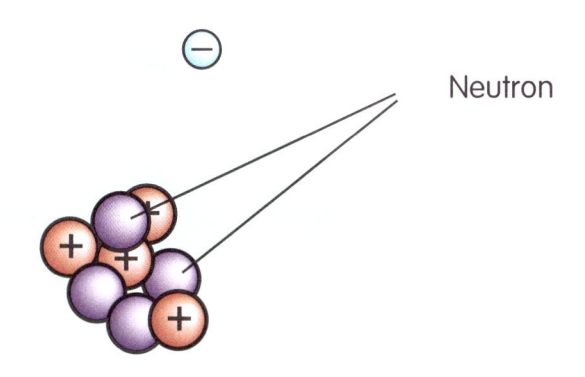

Neutron

Nitrogen
An element made up of seven protons, seven neutrons, and seven electrons.

Nitrogen exists on the Earth naturally as a gas. The Earth's atmosphere is roughly 78% **nitrogen**.

Nonmetal
An element that does not conduct electricity well, is not shiny, and does not bend without cracking.

Some **nonmetal**s you may know are wood, rubber, and plastic.

The prefix "non-" means "not."

Chemistry

Nuclear energy
The energy stored in the nucleus of an atom.

Nuclear energy is stored in a way that is similar to a spring wound very tightly. When an atom's nucleus is split, this energy is released, like when a spring is let go.

Nuclear fission
The process of a heavy nucleus splitting into more than one lighter nuclei.

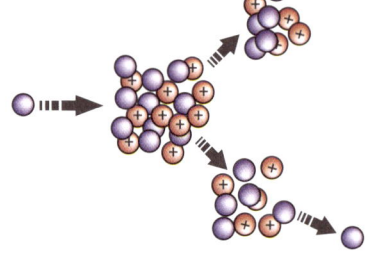

Nuclear fusion
The process of two light-weight nuclei coming together to make one larger, more stable nucleus.

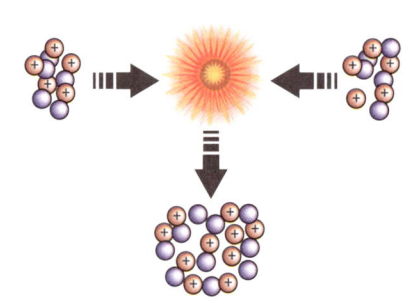

Chemistry

Nuclear reaction
A reaction that involves the nucleus of the atom.

Nuclear fission, nuclear fusion, and radioactive decay are three types of **nuclear reactions**.

Nucleon
A proton or a neutron.

Organic compounds
Most compounds containing carbon.

Organic compounds make up all the structures necessary for life, including proteins, fats, and carbohydrates.

Oxidation number
The number of electrons that would need to be added or removed from an atom in order to change it to its element form.

Hydrogen ions are positively charged because they have lost one electron. Their **oxidation number** is 1 because they would need to gain one electron in order to return to their element form.

Oxidation-reduction reaction
The process by which atoms give or receive electrons.

An atom that gives an electron may become a positively charged ion, while an atom that receives an electron may become a negatively charged ion. This reaction is also called an *electron-transfer reaction* or a *redox reaction*.

Chemistry

Oxygen
An element made up of eight protons, eight neutrons, and eight electrons.

Most plants give off **oxygen**. All animals need **oxygen** in order to survive.

Parts per billion (ppb)
The number of parts of a chemical element found in a *billion* parts of a solid, liquid, or gas.

If you have a billion (1,000,000,000) drops of water in a pool and you add one drop of ink, your pool would have one **part per billion** of ink.

Parts per million (ppm)
The number of parts of a chemical element found in a *million* parts of a solid, liquid, or gas.

If you have a million (1,000,000) drops of water in a pool and you add one drop of ink, your pool would have one **part per million** of ink.

Percent composition
The ratio of each element in a compound.

The **percent composition** of this gold ring is 75% gold, 15% silver, and 10% copper.

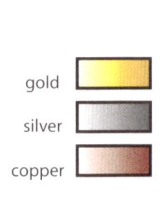

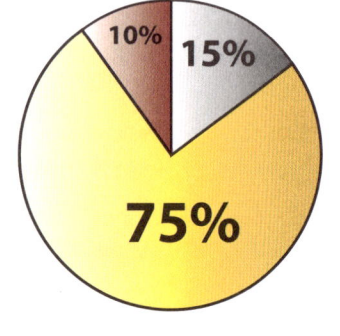

Chemistry

Period
The elements that form a row on the Periodic Table.

The **period** is determined by how many electron shells an element has.

Periodic Table
The chart used by scientists to group different atoms by ways that they are similar.

Please see the **Periodic Table** on page 128.

pH
The measurement of how basic or acidic a solution is.

Scientists measure **pH** on a scale from 1 to 14. 1 is the most acidic; 14 is the most basic.

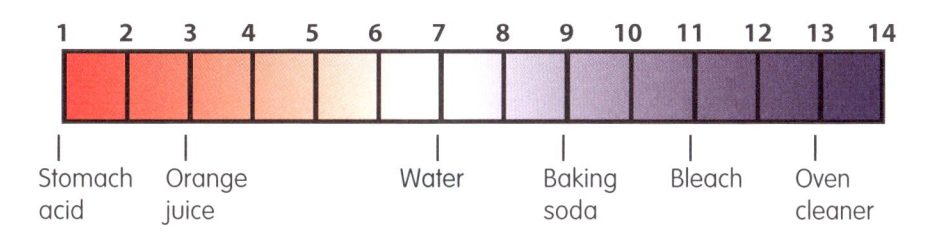

| 1 | 2 | 3 | 4 | 5 | 6 | 7 | 8 | 9 | 10 | 11 | 12 | 13 | 14 |

Stomach acid — Orange juice — Water — Baking soda — Bleach — Oven cleaner

Physical property
A property of a substance that does not change when it is observed.

The weight of a candle is a **physical property**. It does not change when it is measured. The ability of a candle to burn is *not* a **physical property**. You have to change the candle in order to learn whether or not it can burn.

69

Chemistry

Polymer

A large molecule made of many repeating small units that are connected by covalent bonds.

The **polymer** polyethene is a long chain of carbon atoms each bonded to two hydrogen atoms.

The prefix "poly-" means "many."

Precipitate

A solid compound that is made by a chemical reaction in a solution. The **precipitate** falls out of the liquid solution as a solid.

Product

The result of a chemical change.

When iron is exposed to oxygen in the air, a chemical change happens. The **product** of this chemical change is seen as rust.

Property

The qualities of a thing that make it different from other things.

Properties include physical properties like size as well as chemical properties like ability to burn.

Chemistry

Proton
A particle found in the nucleus of an atom. **Protons** have a positive charge and a weight of 1 atomic mass unit (1 amu).

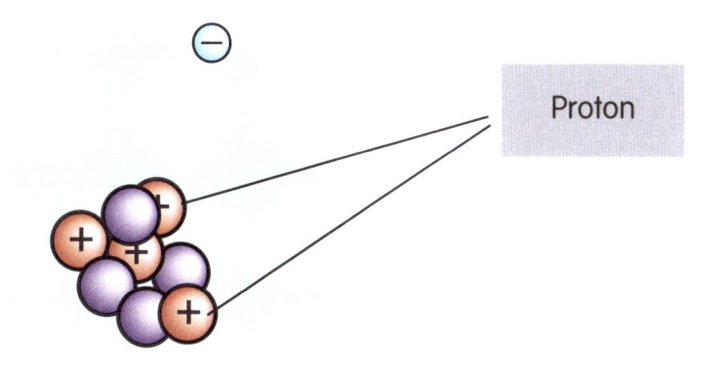

Pure substance
A substance made of one or more molecules that has a specific composition.

A **pure substance** is different from a mixture. In a **pure substance**, every sample will have exactly the same composition. Elements and compounds are **pure substances**.

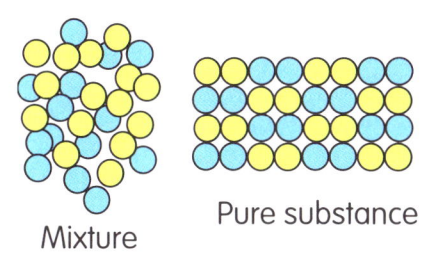

Mixture

Pure substance

Qualitative data
Information expressed in qualities or characteristics, usually found using the five senses.

The *color* of bubbles caused by a reaction is an example of **qualitative data**.

71

Chemistry

Quantitative data
Information expressed in numbers or amounts, usually found using a tool or measuring instrument.

The *number of seconds* it takes for a solid to dissolve in a liquid is an example of **quantitative data**.

Radiation
The giving-off of particles or photons by radioactive atoms during radioactive decay.

An atom that gives off **radiation** can be called *radioactive*.

Radiometric dating
A system for measuring the age of an object by measuring the amount of radioactive decay that has occurred.

Carbon-14 has a half-life of about 5,730 years. Using **radiometric dating**, scientists can figure out that if an object has lost half of its carbon-14, it must be about 5,730 years old.

Radioactive decay
The slow decay of a nucleus into a smaller, more stable nucleus. When this decay happens, atomic particles and/or electromagnetic radiation are given off.

The diagram below shows one carbon-14 atom decaying into one nitrogen-14 atom plus one electron.

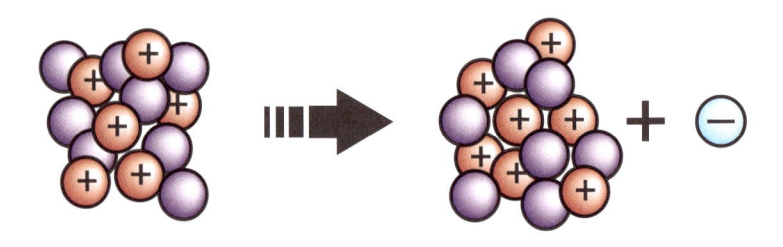

Chemistry

Reactant
A substance that reacts in a chemical change.

In the chemical reaction below, H_2 and O_2 are the **reactants**. They react to form H_2O.
$$2H_2 + O_2 \longrightarrow 2H_2O$$

Redox reaction
Please see *oxidation-reduction reaction*.

Saturated solution
A solution that is holding the most dissolved solute that it can at a given temperature. If any more solute is added, it will not be able to dissolve.

If a glass of water is saturated with salt, this means that the water can not hold any more salt in solution. If you add more salt to the glass, the salt will collect at the bottom of the glass as a solid.

Solubility
The measure of how much of a substance can dissolve in a given amount of another substance.

Most alcohols dissolve easily in water. Therefore, alcohol has a high **solubility** in water. Oil does not dissolve in water at all. Therefore, oil has a very low **solubility** in water.

Solute
The substance dissolved in a solution.

In this diagram, a student pours 50 ml of food coloring into 1L of water. The food coloring is the **solute**.

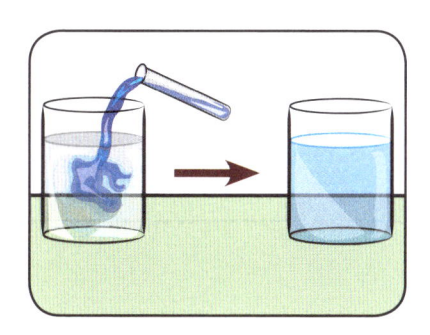

Chemistry

Solution
A mixture of two or more substances.

Chocolate milk is a **solution** made up of a solvent (milk) and one solute (chocolate syrup).

Solvent
The substance that dissolves a solute.

In the diagram below, a student pours 50 ml of food coloring into 1L of water. The water is the **solvent**.

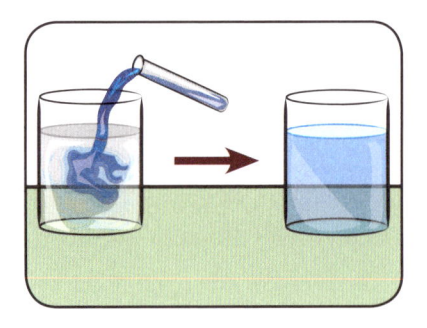

Specific heat
The amount of heat energy needed to raise the temperature of one gram of substance by one degree Celsius (1 °C) or one kelvin (1 K).

Different substances have different **specific heats**. Iron has a lower **specific heat** than water, which means it takes less energy to raise the temperature of iron than it does to raise the temperature of water.

State of matter
The form that a substance is in.

There are three major **states of matter**: solid, liquid, and gas.

Chemistry

Subatomic particles
Particles that are smaller than an atom and may make up an atom itself.

Protons, neutrons, and electrons are all **subatomic particles**.

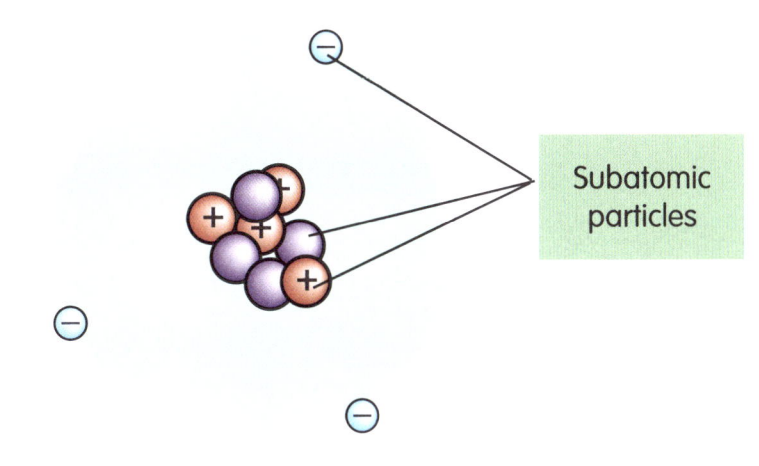

Subatomic particles

Titration
A test used to determine how acidic or basic a solution is.

Performing a **titration** involves adding a basic solute with a known pH to an acidic solvent with an unknown pH. The solute is added until the solution reaches a pH of 7. A **titration** can also be reversed, using a known acidic solute to determine the pH of an unknown base.

Unsaturated solution
A solution that is holding less than the maximum amount of a given solute.

If a glass of salt water is an **unsaturated solution**, this means the water is still able to dissolve more salt in it.

Chemistry

Valence electron
An electron that is available to be bonded, gained, or lost to another atom.

Valence electrons are the electrons in an atom's outermost shell. These electrons determine how an atom will react with another atom.

Vaporization
To change from a liquid to a gas with or without boiling.

Water goes through **vaporization** as it moves from liquid to gas form. You can see this happening as puddles dry up after a rainstorm.

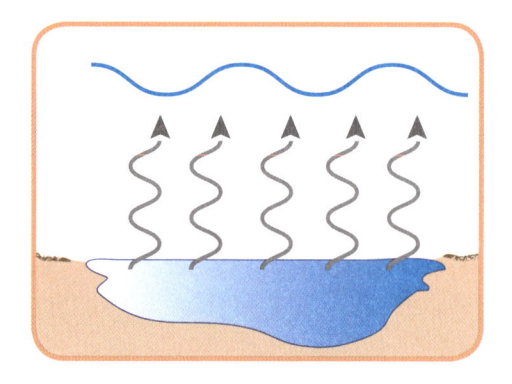

Physical Science

Absolute zero
The temperature at which atoms stop moving entirely. It is measured as -273° Celsius or 0 Kelvin.

Absorb
To take in something.

A sponge can **absorb** liquids.

AC circuit
A circuit in which the direction of the current changes.

AC stands for *alternating current*.

Accelerate
To speed up.

To win the bicycle race, Dina had to **accelerate** by pedaling faster.

Alloy
A mixture of elements that have metallic properties.

Jewelry is often made of gold **alloys**, or gold mixed with copper, zinc, or nickel.

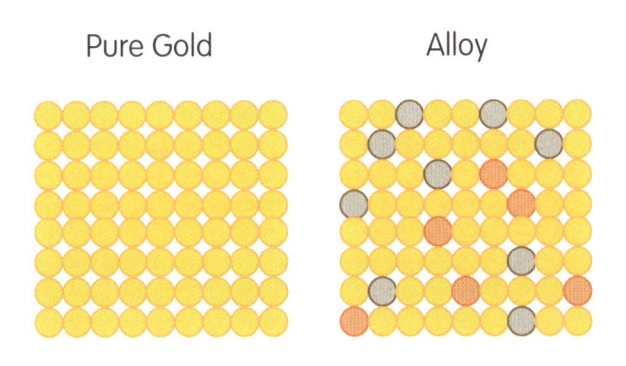

Pure Gold   Alloy

77

Physical Science

Ampere (A)
The unit of measurement for electrical current.

The tool used to measure **amperes** is called an *ammeter*.

Amplitude
The height of an electromagnetic wave, or half the distance from crest to trough.

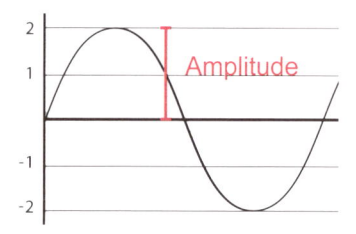

The **amplitude** of the wave in this picture is 2.

Axis (plural: axes)
Imaginary lines that help describe the movement of objects in space. These lines are commonly called the *x-axis*, *y-axis*, and *z-axis*.

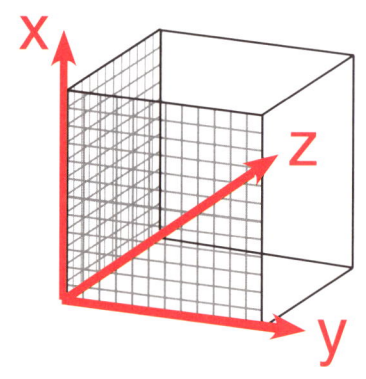

The **axes** refer to movement that is up/down (x-axis), left/right (y-axis), and forward/backward (z-axis).

78

Physical Science

Centripetal force
The force directed toward the center of a circular path.

If you are swinging a string with a ball on the end around in a circle, the force of the string pulling the ball towards the center of the circle is the **centripetal force**.

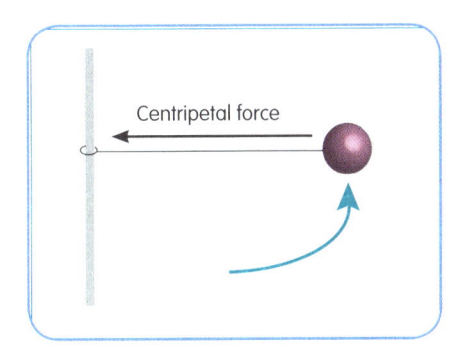

Charge
A basic electrical characteristic of matter.

Charge is shown in two forms: positive and negative.

Circuit
A path that begins and ends in the same place.

Physical Science

Color
The appearance of an object based on which parts of the visible light spectrum are absorbed and which are reflected.

A red ball absorbs all parts of the visible light spectrum except red. The ball reflects red light, so you see it as a red ball.

Convection
The movement of heat through a fluid, such as water or air.

Convection ovens rely on heated air to flow around and cook food.

Coulomb (C)
The unit of measurement for electric charge.

Crest
The highest point of a wave.

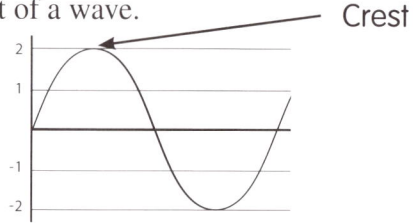

Physical Science

Current
A flow of something from one place to another.

Electricity flows in a **current**. Water can also flow in a **current**.

DC circuit
A circuit in which current flows in only one direction, usually from negative (-) to positive (+).

DC stands for *direct current*.

Electric charge
A basic electrical characteristic of matter.

An electron's **electric charge** is -1. A proton's **electric charge** is +1.

Electric circuit
A path along which electrons can flow.

An **electric circuit**, like all circuits, begins and ends at the same point.

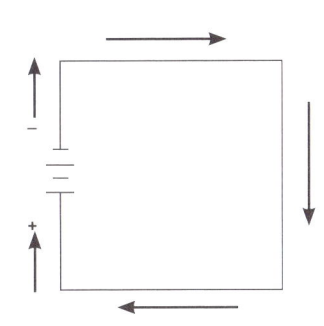

Electric conductor
A substance through which electrons can move easily.

Wires are often made of copper because copper is a very good **electric conductor**.

Physical Science

Electric field
An area where an electric force or charge can be detected.

Food in a microwave cooks because of the **electric field** inside the appliance.

Electric insulator
A substance that stops electrons from moving to one place from another.

Rubber, glass, and plastic are all **electric insulators**. This is why many wires are coated with plastic.

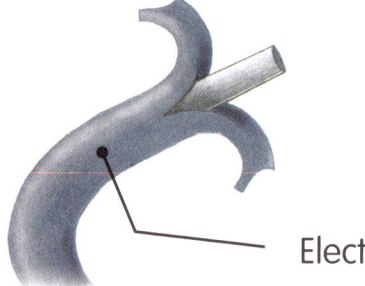

Electric insulator

Electricity
The movement of electrons.

In a home, **electricity** is the movement of electrons along wires. The **electricity** transports energy that is used to power anything that is connected to the electric circuit, such as televisions and lamps.

Physical Science

Electromagnetic radiation

A form of energy that travels through space as a wave.

Electromagnetic radiation includes visible light as well as radio waves and x-rays. All the forms of **electromagnetic radiation** together are called the *electromagnetic spectrum.*

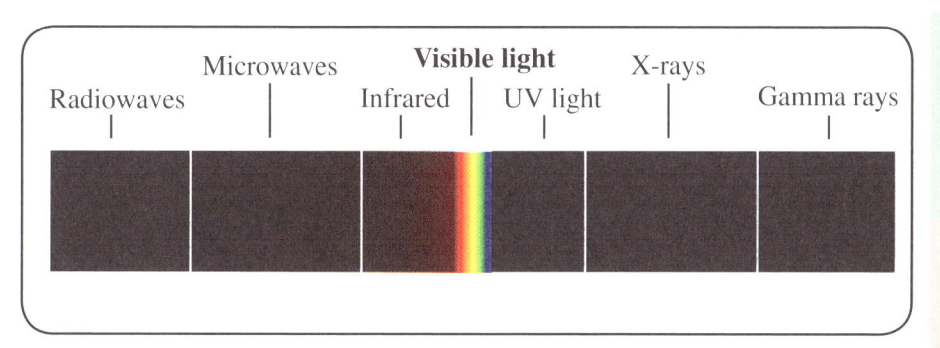

Emission spectrum

A diagram or graph that illustrates the wavelengths of energy that a substance gives off.

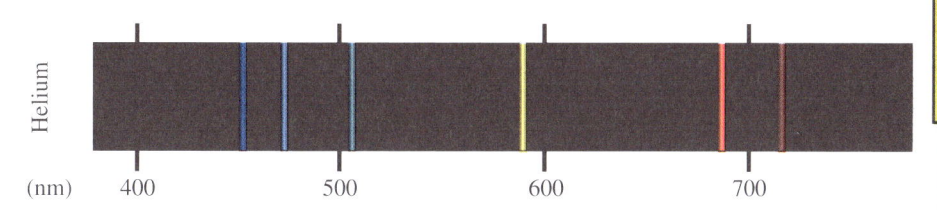

Fission

The process by which the nucleus of a large atom, such as uranium, splits into smaller nuclei.

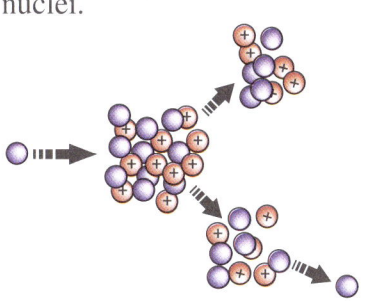

83

Physical Science

Force
A push or a pull.

Opening a door takes **force**: it is a pull in one direction.

Frequency
The measurement of the number of wave crests that pass a point in one second.

In this picture, the wave on top has a lower **frequency** than the wave on the bottom.

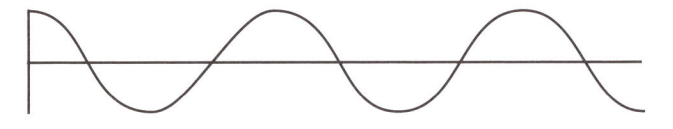

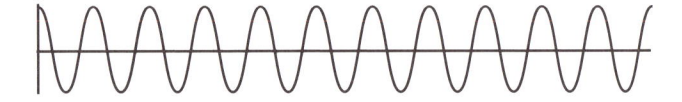

Friction
A force that slows down or stops an object.

A ball rolling on carpet will eventually stop rolling. This is because of the **friction** between the carpet and the ball.

Fusion
The combining of the nuclei of two atoms.

When nuclei come together during **fusion**, energy is released.

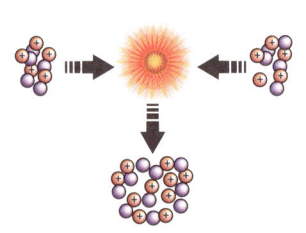

Physical Science

Gravitational force
The force of attraction between two pieces of matter.

All matter has some **gravitational force**, but it is much easier to measure this force when the piece of matter is very large. The Earth has a much stronger **gravitational force** than a human being does.

Gravitational potential energy
Energy that is gained when an object is lifted.

Half-life
The time it takes for half of all of one particular kind of nucleus to decay.

Carbon-14 has a **half-life** of 5,730 years. In 5,730 years, there will be half the amount of carbon-14 in the world than there is today.

Heat
The energy transferred from one object to another object that has a different temperature.

Heat always moves from an area of higher temperature to an area of lower temperature. **Heat** will continue to move in this direction until both objects are the same temperature.

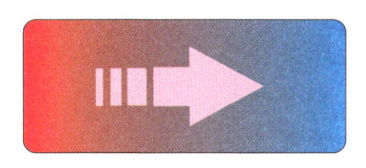

Heat conductor
A substance through which heat can move easily.

Metals are excellent **heat conductors**.

Physical Science

Heat insulator
A substance that stops heat from moving from one place to another.

Fiberglass, wool, and plastic foam are all **heat insulators**.

Joule (J)
The unit of measurement for energy.

A **joule** is equal to the amount of energy it takes to exert the force of one newton for a distance of one meter. The **joule** also measures the energy required to move one coulomb of electrical charge through an electric potential difference of one volt.

Kinematics
The study of the motion of objects.

Equations that are used in the study of **kinematics** are known as *kinematic equations*.

Kinetic energy
The energy an object has because of its motion.

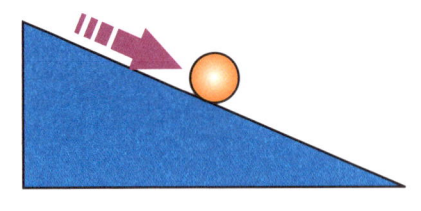

This ball is rolling. It has **kinetic energy**.

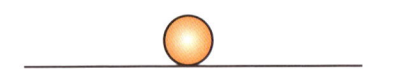

This ball is sitting still. It has no **kinetic energy**.

86

Physical Science

Lens
A piece of glass or plastic that focuses or spreads out light rays.

Lenses can be used to focus light in or spread it out.

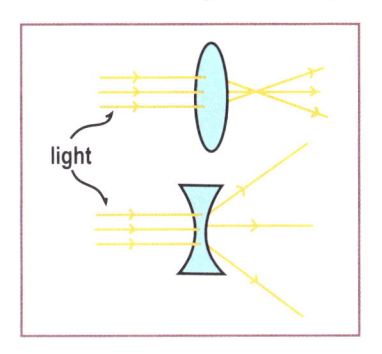

Magnet
An object that can attract iron.

Any object that acts like a **magnet** can be called *magnetic*.

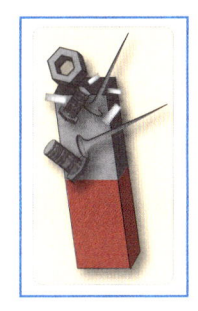

Magnetic field
The places where the force of a magnet can be felt.

The lines in the picture show the **magnetic field** of this magnet.

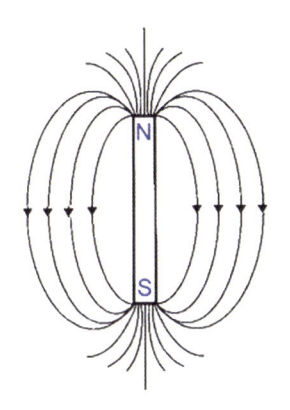

87

Physical Science

Magnetic poles
The ends of a magnet that have the strongest pull.

Magnetic poles can be either positively or negatively charged.

Mechanical energy
The total of the kinetic energy and all of the potential energies added together.

In physics, **mechanical energy** can be written as an equation: $KE + PE = ME$, where KE = kinetic energy and PE = potential energy.

Momentum
A measure of movement.

Momentum = mass × velocity.

Motion
A change in the position of an object over time.

Physical Science

Newton's Laws
The first accurate group of laws that explained the movement of things like spinning objects and projectiles on Earth.

> **Newton's Laws**
> - Law of Inertia
> - Law of Motion
> - Law of Reciprocal Actions
> - Law of Universal Gravitation
>
> **Newton's laws** are also called *classical physics*. These laws are named for scientist Sir Isaac Newton, who discovered many of the laws of physics in the 17th century.

One-dimensional motion
Moving along only one axis of space.

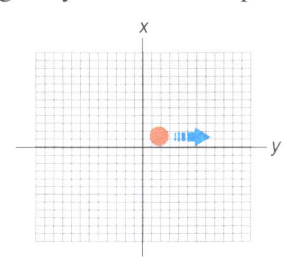

One-dimensional collision
When two objects moving along the same line run into each other, or collide.

Two trains facing each other and running on the same track will eventually have a **one-dimensional collision**.

89

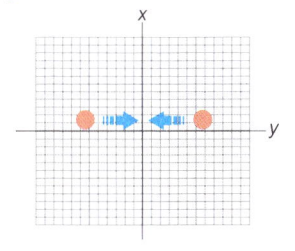

Physical Science

Optical system
Two or more lenses or mirrors that work together to do a job.

A telescope is a type of **optical system**.

Parallel circuit
An electric circuit where each item in the circuit is connected directly to the electrical source.

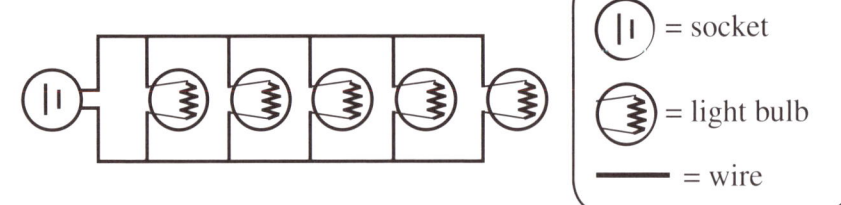

= socket
= light bulb
= wire

Polarization
The process where a molecule or object develops areas of positive charge and areas of negative charge.

Anything that shows **polarization** is said to be *polarized*. Clouds become polarized because the water droplets within them rub together. Positive charges move to the top of the cloud while negative charges move to the bottom. The **polarization** of clouds is what causes lightning.

Physical Science

Position
The location of an object.

You can express the **position** of an object using an *x and y axis.* The **position** of the dot on the axis below is (2, 3).

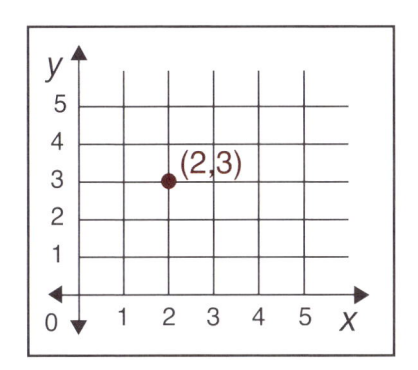

Potential energy
The energy that something has because of its position or shape.

The ball on the ramp has **potential energy.** If it were to begin rolling down the ramp, it would lose its **potential energy** and gain kinetic energy.

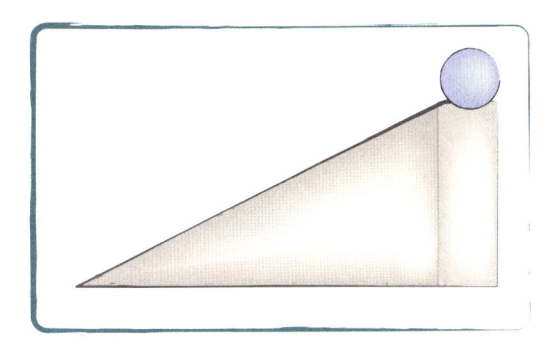

Power
The measure of how fast work can be done.

It takes more **power** for a horse to pull two logs in the same amount of time than it does to pull one log.

Physical Science

Pressure
The amount of force that pushes against a certain area.

Pressure = force ÷ area.

Reflect
To throw back light or another electromagnetic wave.

A mirror will **reflect** light.

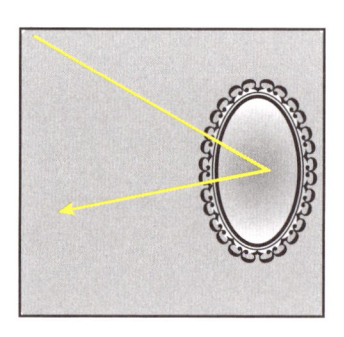

Refract
To bend light or change its path.

Light will **refract** when it travels through air into water.

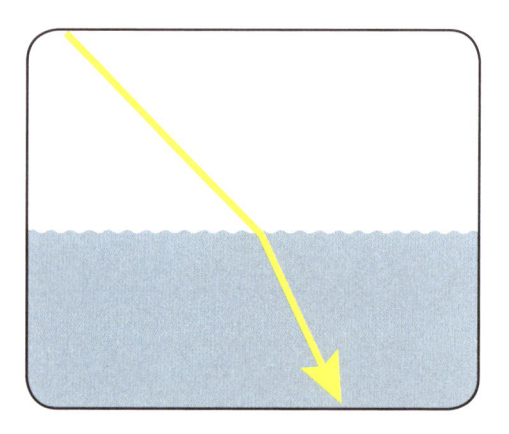

92

Physical Science

Resistance
Something that works against the flow of an electric current.

Any item that provides **resistance** to a current is called a *resistor*.

Scalar
A quantity that has a number measurement, but no direction measurement.

Temperature is a **scalar** quantity. Temperature is measured in numbers, but not by direction.

Semiconductor
A substance that can act like a conductor and an insulator.

Series circuit
An electric circuit that has one continuous loop and each item in the circuit receives the same amount of current.

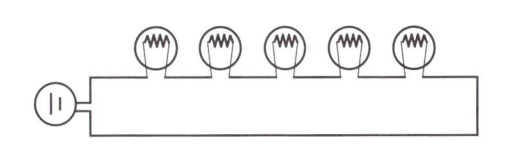

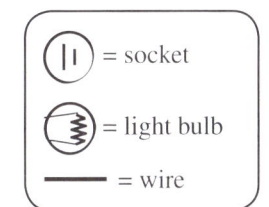

= socket
= light bulb
= wire

Simple machine
A device that is made of one or two parts and only needs one application of force to do work.

There are six kinds of **simple machines**:

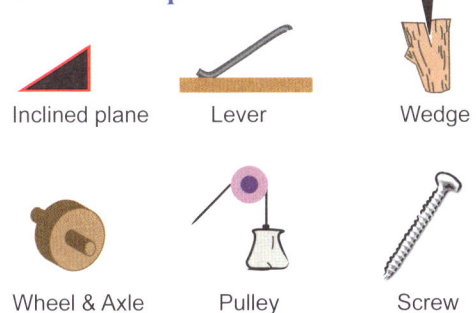

Inclined plane   Lever   Wedge

Wheel & Axle   Pulley   Screw

93

Physical Science

Sound
A set of vibrations that travel as waves through air and other substances. When this set of vibrations reaches your ear, you can hear them.

Some **sound** waves are too high or too low for humans to hear. There are animals that can hear these types of **sounds**. For example, elephants hear very low **sounds** and dogs can hear very high **sounds**.

Stationary charges
Positive (+) and negative (-) charges that are not moving.

Tesla (T)
A unit of measurement for the strength of a magnet.

A small magnet has the strength of about 0.01 **tesla**.

The **tesla** was named after a scientist named Nikola Tesla, who spent his lifetime studying electricity and magnetism.

Thermal energy
The energy of an object based on the amount of heat it holds.

If two objects at different temperatures are placed together, the object with more heat will transfer **thermal energy** to the object with less heat until they reach equilibrium. Temperature is a measure of **thermal energy**.

Physical Science

Thermodynamics
The study of how energy moves within systems.

Thermodynamics involves the study of light energy, heat energy, and many other forms of energy.

The rules for all forms of energy are known as the *laws of thermodynamics*.
- First law: energy cannot be created or destroyed
- Second law: Entropy is more likely to increase than decrease in systems
- Third law: As temperature approaches absolute zero, the energy of a system approaches a constant

Three-dimensional motion
Moving along the X, Y, and Z axes.

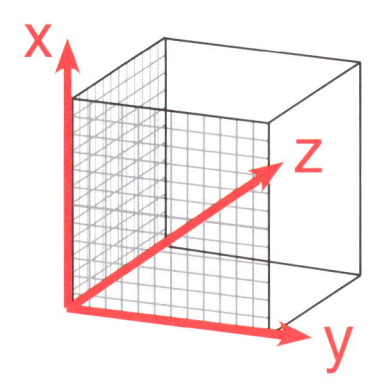

Torque
A measure of the force it takes to rotate an object around an axis.

95

Physical Science

Trough
The lowest point on a wave.

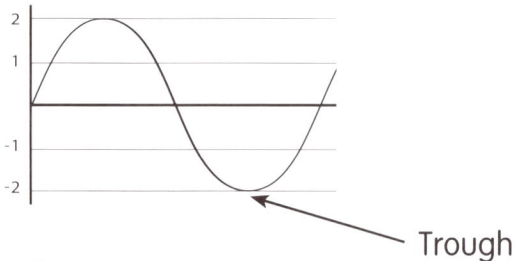

Trough

Two-dimensional motion
Moving along the X and Y axes.

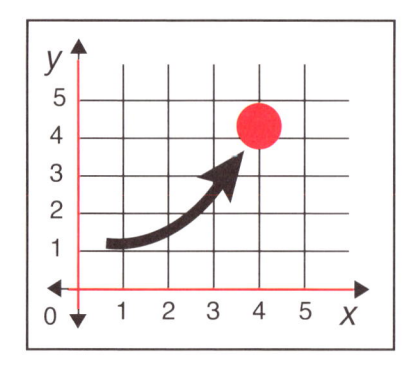

Units
A set amount of something that is used for measurement.

Some **units** include:

Inches	Length
Cup	Capacity
Ounce	Weight

Vector
A physical quantity that has both magnitude and direction.

40 mph is a **vector** quantity. It has both speed (40 mph) and direction (east).

Physical Science

Vibration
The back-and-forth movement of an object.

Many cell phones use **vibration** as a way to let you know someone is calling.

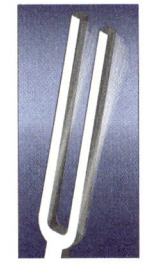

Volt (V)
The unit that measures how much energy there is per coulomb of charge.

The **volt** is named after 18th century physicist Alessandro Volta, who developed the first electric battery.

Voltmeter
A tool that measures the number of volts across a part of an electric circuit.

Watt
The measurement of the amount of energy in an electric circuit.

A 100-**watt** light bulb will typically give off more light than a 75-**watt** light bulb.

In physics, **watts** are often expressed in terms of the amount of watts of energy used in a given amount of time, as in one kilowatt-hour, or kwh.

Physical Science

Wave
A disturbance in matter or space.

Waves move energy from one place to another.

Wavelength
The distance from one wave crest to the next wave crest.

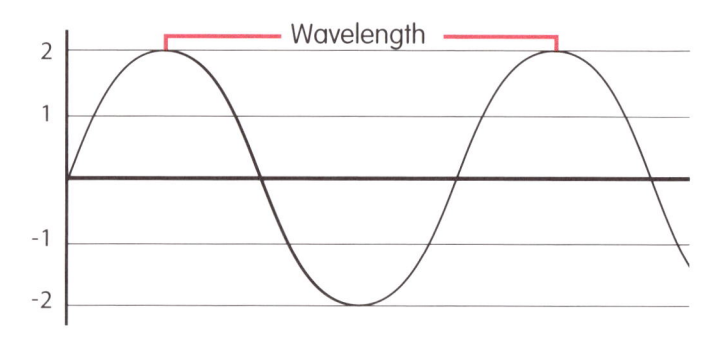

Work
A measure of how much energy it takes to move an object.

Work = force × displacement.

Earth and Space

Acid rain
Rain that has a low pH due to pollutants such as automobile exhaust.

Acid rain can hurt plants and animals. It can also damage buildings.

Aquifer
An underground layer of earth or stone that holds water.

Asteroid
A large rock in space.

Most **asteroids** travel around the Sun between Mars and Jupiter.

Asthenosphere
The lower part of the Earth's mantle. It is made of solid rock that is under so much pressure, it begins to flow like a liquid.

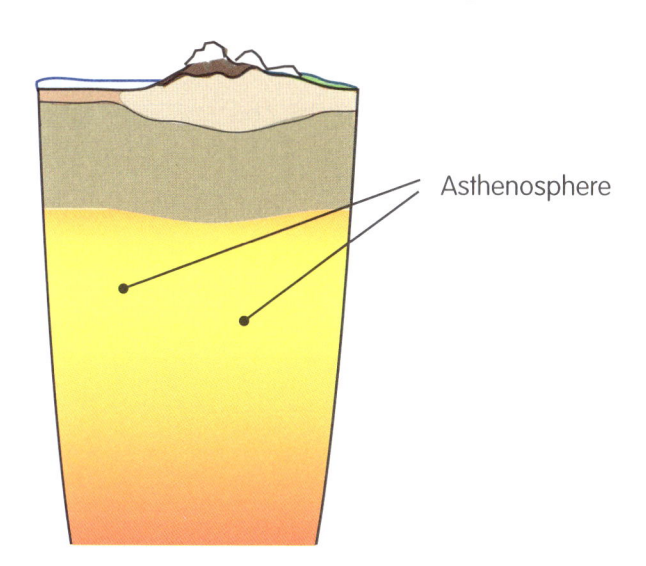

Asthenosphere

99

Earth and Space

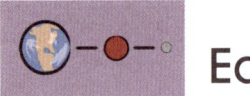

Astronomical unit (AU)
A unit used to measure distance in the solar system.

One **astronomical unit** is equal to the distance from the Earth to the Sun, which is about 150 million kilometers or 93 million miles.

Astronomy
The study of objects outside Earth's atmosphere.

A scientist who studies **astronomy** is called an *astronomer*.

Atmosphere
A thin layer of gas that surrounds a planet.

Earth's **atmosphere** protects us from the harmful rays of the Sun.

Atmospheric gases
The gases that are found in the atmosphere.

Atmospheric gases include oxygen, nitrogen, and carbon dioxide.

Autumnal equinox
One of two days of the year when the Sun is directly overhead at the Equator (the other day is called the *vernal equinox*). On this day, there are an equal number of hours of daytime and nighttime in both the Northern and Southern hemispheres.

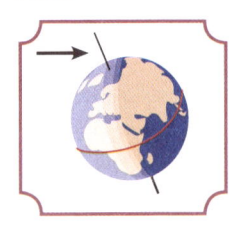

Axis
An imaginary line that runs through the center of a planet.

The planet's **axis** is the imaginary line that the planet revolves around.

Earth and Space

Big Bang theory
A theory that explains how the universe might have been created.

In the **Big Bang theory**, the universe started out as a single point and has been expanding from this one point ever since.

Biosphere
All of the places on Earth where there is life.

Black hole
A dense area where a star used to be. This area is so dense that its gravity pulls in everything, including light.

Carbon cycle
The path that carbon atoms take as they move through different parts of the Earth.

Carbon moves through living things (plants and animals) and through nonliving things (water, air, and rock) in the **carbon cycle**.

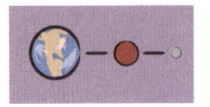

Earth and Space

Cleavage
The ability of a mineral to break easily along a flat plane.

Climate
The kind of weather an area has most of the time.

The southwestern United States has mainly a hot and dry **climate**.

Clouds
A body of tiny water droplets suspended in the atmosphere.

There are many different kinds of **clouds**. The altitude and composition of a **cloud** determine what kind it is.

Some types of clouds
- Cirrus
- Cumulus
- Cumulonimbus
- Stratus

Comet
A piece of ice, dust, and rock that orbits the Sun.

Comets have tails. These tails always point away from the Sun.

Earth and Space

Continent
A large land mass on the Earth.

Continental drift
A theory that all the Earth's land masses are moving very slowly with time.

All the Earth's land masses used to be one large land mass millions of years ago. This large land mass then broke into smaller pieces. The pieces slowly moved away from each other into the places where they are today.

Convection
Energy that is transferred by the flow of a heated substance.

Convection ovens are ovens that rely on heated air to flow around and cook food.

Core
The layer of material at the center of the Earth.

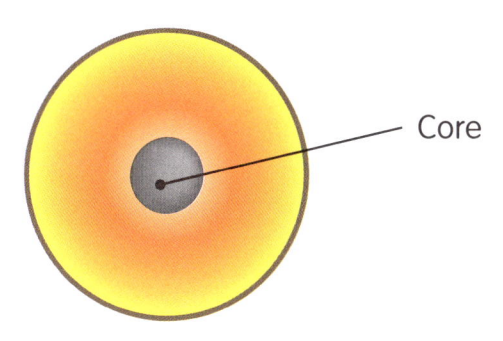

103

Earth and Space

Crust
The outer brittle, rocky surface of the Earth.

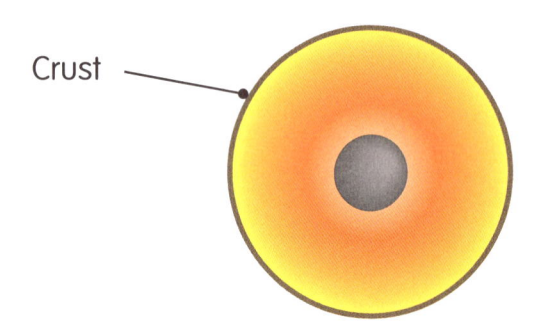

Current
Part of a body of air or water that is moving in a single direction.

Rivers flow in a **current**.

Delta
A large deposit of silt and clay that forms when a river meets a large body of water such as a lake or ocean.

The Nile River **delta** is the area where the Nile flows into the Mediterranean Sea.

Deposition
The settling of material out of the air or water.

Oceans *deposit* shells and pebbles on the beach in a process called **deposition**.

Earth and Space

Drought
A long time with no rain in an area that usually has rainfall.

During a **drought**, plants can become damaged and even die.

Earthquake
Shaking ground that is caused by the Earth's crust moving.

In 1923, a major **earthquake** hit Tokyo, the capital city of Japan. The Tokyo **earthquake** killed nearly 100,000 people.

Eclipse
A shadow cast by one object onto another object in space.

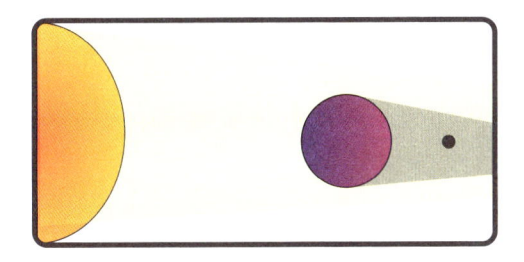

Equator
The imaginary line that goes around the Earth. This line is 0° latitude. It divides the Earth into the northern and southern hemispheres.

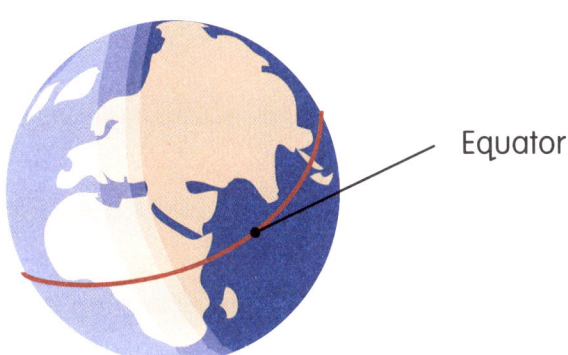

Equator

105

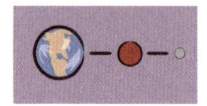

Earth and Space

Erosion
The movement of particles that have been worn away by weathering.

Because of **erosion**, some mountains are not as tall as they used to be.

Fault
A large break in the Earth's crust. Different pieces of the Earth's crust move closer to each other or farther apart along these **faults**.

Earthquakes often occur along **fault** lines because they are areas of frequent movement of the Earth's crust.

Flood
Water spilling over the sides of a stream or river, or flowing past the normal shoreline of an ocean. This is usually caused by a large amount of rainfall or high winds.

Floods can be harmful to a populated area, but they can also help the environment.

Fossil
Evidence of a living thing that died long ago.

When something turns into a **fossil**, it becomes *fossilized*.

Earth and Space

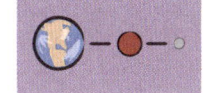

Fossil fuel
Fuels that are formed from the remains of plants and animals that died millions of years ago.

Coal is one type of **fossil fuel**.

Freshwater
Water that is not salty.

You can find **freshwater** in many lakes and almost all streams.

Fuel
Material that is burned to create heat or power.

Some types of **fuel** you may use are gasoline, coal, and wood.

Galaxy
A group of millions of stars in space.

The Earth is part of a **galaxy** called the *Milky Way*.

Geographic poles
The points where the Earth's axis passes through its surface.

There are two **geographic poles**: the North Pole and the South Pole.

107

Earth and Space

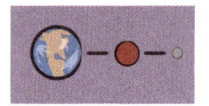

Glaciation
A geological period in which huge sheets of ice from the poles grow and extend toward the equator.

Glaciation is what happens when the Earth enters an ice age.

Glacier
A large body of ice that moves downhill very slowly.

Global warming
A rise in the average temperature of a planet.

Because **global warming** is heating up the Earth, the polar ice caps may slowly warm and begin to melt. This could affect many arctic animals, such as foxes and polar bears.

Gravity
The force that pulls two objects toward each other.

Gravity keeps people from floating off the Earth. It also keeps the Earth from floating away from the Sun.

Isaac Newton was the first scientist to figure out what are now known as the "laws of **gravity**." One of Newton's most important discoveries is that all objects are pulled toward the Earth at the same rate, regardless of their mass.

Earth and Space

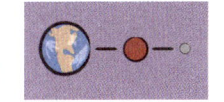

Greenhouse effect

Earth's atmosphere traps light and heat from the Sun in the same way a greenhouse does.

On Earth, the **greenhouse effect** is increasing and causing global climate change.

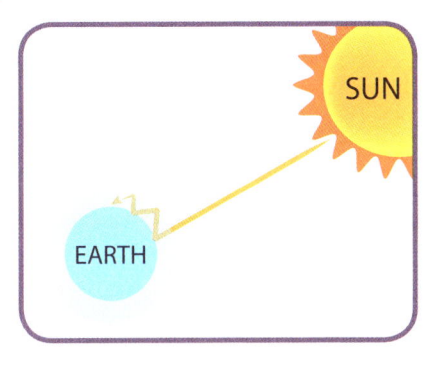

Greenhouse gases

The atmospheric gases that add to the greenhouse effect.

Greenhouse gases include water vapor, carbon dioxide, and ozone.

Groundwater

Water held in aquifers under the Earth's surface.

Groundwater is often the only place where people can get fresh water.

Half-life

The time it takes for half of a sample of a radioactive element to decay.

Carbon-14 has a **half-life** of about 5,730 years. This means that for a 1 ounce sample of carbon-14, in 5,730 years, half of that sample will have decayed into another substance, leaving only ½ an ounce of carbon-14.

Earth and Space

Hertzsprung-Russell Diagram (H-R diagram)
A graph that shows the size, brightness, and temperature of stars.

Ice age
A period of the Earth's history when much of the planet was covered in ice and snow.

The Earth's most recent **ice age** ended about 10,000 years ago.

Igneous rock
A rock that forms when hot liquid rock cools and then hardens. **Igneous rock** can form on the surface of a planet or deep underground.

Kepler's laws
The three laws about the movement of the planets proposed by German astronomer Johannes Kepler in the early 1600s.

Kepler's laws
- The first law states that planets move around a star in an elliptical (oval) orbit.
- The second law states that a planet travels faster when it is closer to a star than when it is farther away.
- The third law states that the farther a planet is from its star, the longer it takes for the planet to move around the star once.

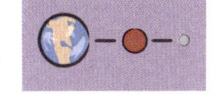

Earth and Space

Lake
A natural or man-made land mass that is full of water.

Lakes can be full of either salt water or freshwater.

Land mass
Any large area of land, such as an island, peninsula, or continent.

The continent of Australia is an example of a **land mass**.

Latitude
A measurement of distance north or south of the Earth's equator.

Latitude is measured in degrees.

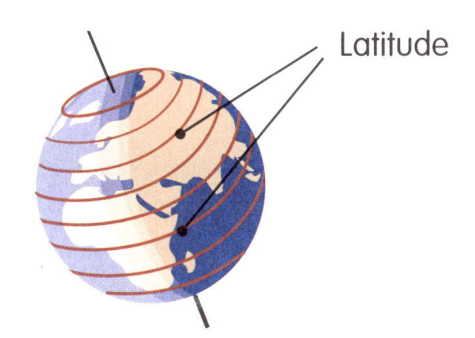

Lava
Hot liquid rock on the surface of a planet.

Earth and Space

Lithosphere
The layer that makes up the outermost part of the Earth. The **lithosphere** is made of the crust and a thin layer of mantle.

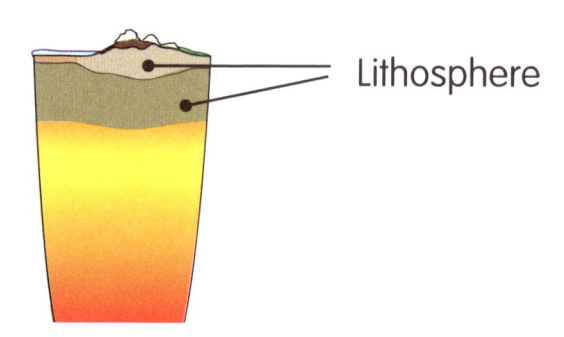

Lithosphere

The prefix "litho-" means "stone."

Longitude
A measurement of distance east or west of the Earth's prime meridian.

Longitude is measured in degrees. 0° **longitude** is called the *Prime Meridian*. This **longitude** line runs through the city of Greenwich, England.

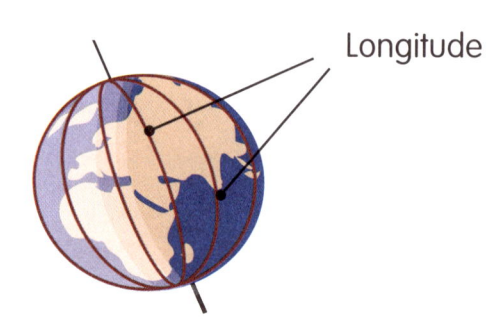

Longitude

Earth and Space

Lunar eclipse
When the Earth passes between the Sun and the Moon. Earth's shadow falls on the Moon and blocks it from view.

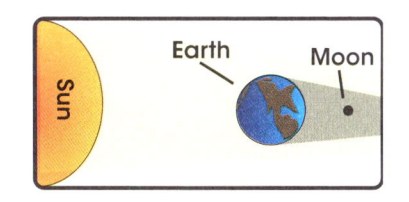

A **lunar eclipse** can occur only at night during a full moon.

Magma
Hot liquid rock below the surface of the planet.

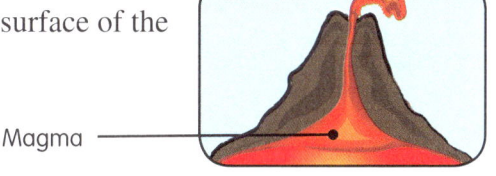

Magma

Mantle
The layer of solid rock between the Earth's crust and its core.

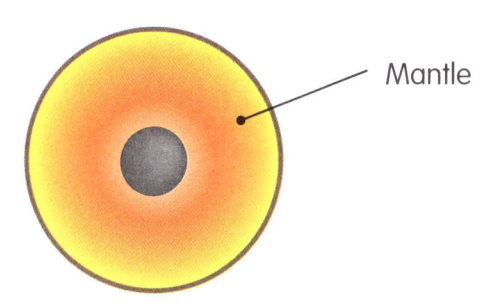

Mantle

Metamorphic rock
A rock that has been changed by heat and pressure.

All **metamorphic rocks** were once igneous or sedimentary rocks. Then, heat or pressure changed them.

Earth and Space

Metamorphose
To change in physical appearance or structure.

Caterpillars **metamorphose** into butterflies.

The term "morph" means "to change into something else."

Meteor
A rock from space that burns up as it enters the Earth's atmosphere.

When many **meteors** fall at once, it is called a meteor shower.

Meteorite
A rock from space that lands on the surface of the Earth.

Meteorites are meteors that do not burn up completely before they reach the Earth's surface.

Meteoroid
A small rock that orbits the Sun or another large body in space.

Meteoroids become meteors when they are drawn into the Earth's atmosphere by gravity.

Earth and Space

Mineral
A naturally formed substance that has a specific chemical composition and crystalline structure.

All elements are **minerals**. Coal, sand, and salt are also **minerals**.

Moon
The natural satellite of a planet.

Earth has one **moon**, called the *Moon*. Jupiter has sixty **moons**. The four largest ones are called Io, Europa, Ganymede, and Callisto.

Mountain
A high land mass rising from the Earth's surface.

Nebula
A huge cloud of hot gas and dust in space. This cloud can collapse into itself and form a new star.

Earth and Space

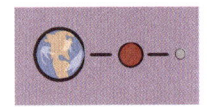

Ocean
The whole body of salty water covering more than two-thirds of the Earth's surface.

The large body of water between Japan and the western United States is known as the Pacific **Ocean**.

Orbit
The path an object takes around another object in space.

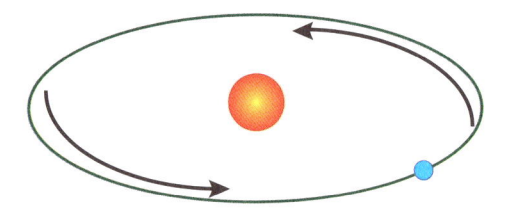

The **orbit** of the Earth is almost circular.

This word can also mean the action of going around another object in space.

The Earth **orbits** the Sun.

Ore
A mineral or rock that holds one or more useful metals.

Iron **ore** is a rock that contains iron along with other substances.

Ozone
A molecule made of three oxygen atoms. **Ozone** is a gas.

Ozone is found mainly in a part of the upper atmosphere called the *ozone layer*. Here, it protects the living things on Earth by absorbing harmful radiation. When **ozone** is produced in the lower atmosphere (the air we breathe), it can be harmful to living things.

Earth and Space

Ozone depletion
The result of the human use of chemicals that break down the ozone layer in the Earth's atmosphere.

Ozone depletion damages the Earth because without ozone, the planet is not protected from the Sun's ultraviolet rays.

Pangaea
The ancient land mass that was made up of all the present-day continents joined together.

On a map, it looks as if South America and Africa could join together like pieces of a puzzle. According to some scientists, many millions of years ago, these land masses were joined in one large mass called **Pangaea**.

Planet
A large ball of rock or gas that orbits a star.

Saturn is one of at least eight **planets** that orbit the Sun.

Planetary system
A group of objects in space containing at least one star and other orbiting objects, such as planets, moons, and asteroids.

The **planetary system** to which the Earth belongs is called the *Solar System*.

Earth and Space

Plate
One of the sections of the Earth's lithosphere. There are about twelve plates that make up the Earth's surface. They are sometimes called *tectonic plates*.

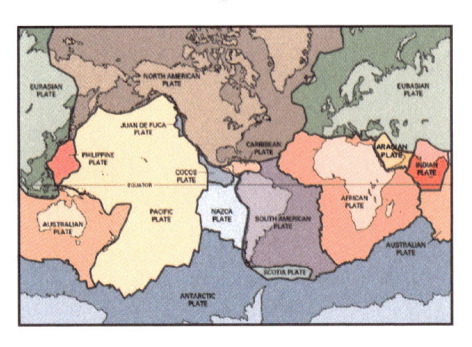

Plate tectonic theory
The theory that says the Earth's crust is broken into plates that move very slowly over millions of years. Some of these plates are moving into one another, while others are moving away from one another.

Prime Meridian
An imaginary line running from the North Pole, through Greenwich, England, to the South Pole over Earth's surface.

The **Prime Meridian** is 0° longitude.

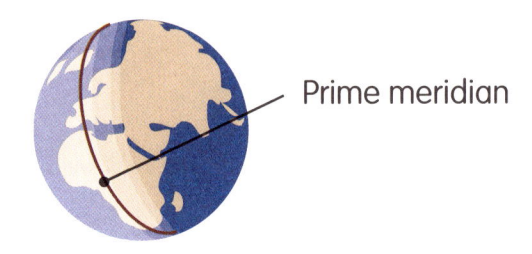

Prime meridian

Radiation
Energy that comes to the Earth from space.

Most of the **radiation** directed at the Earth is absorbed by Earth's atmosphere.

Earth and Space

Radioactive decay
The process by which an element releases radiation and particles and becomes a different element.

Scientists can measure **radioactive decay** of elements and use this information to figure out the approximate age of structures on the Earth. This process is called *radiometric dating*.

Rock
A hard object made of minerals.

The three types of **rock**
- igneous
- metamorphic
- sedimentary

Rock cycle
The process that rocks go through as they change from one kind of rock to another.

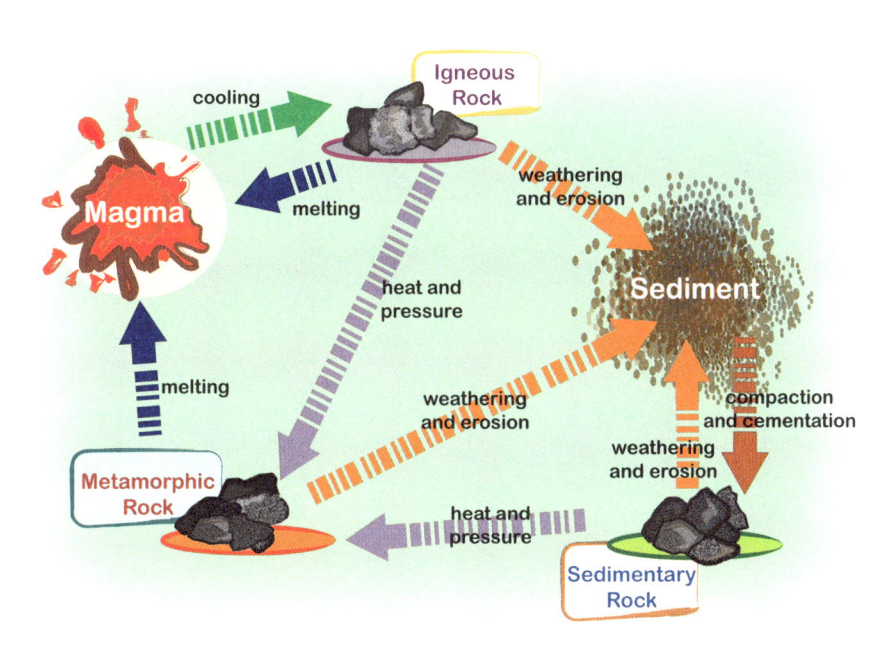

119

Earth and Space

Rotation

The movement of a planet turning around its axis.

A planet undergoing **rotation** is said to *rotate*. Planets rotate about an axis, and they revolve around a star.

Satellite

An object that travels around a larger object in space.

The Moon is a natural **satellite** of the Earth. The Earth also has many man-made **satellites** that help us to communicate.

Sea-floor spreading

A process by which new ocean crust is formed at mid-ocean ridges where the plates move apart. At these ridges, small amounts of magma come to the surface, cool, and become new crust.

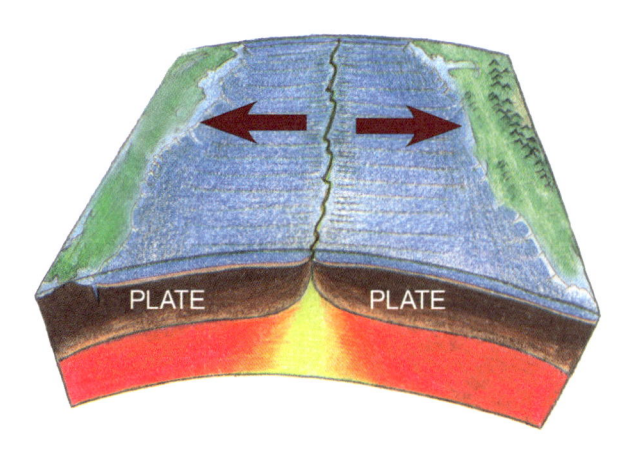

Sediment

Material deposited on the Earth's surface by weathering and erosion.

Sediment can eventually form into sedimentary rock.

Earth and Space

Sedimentary rock

A rock made from pieces of other rock. Layers of sediment that are under pressure for long periods of time can solidify into rock.

Sand on the beach → Millions of years → Rock

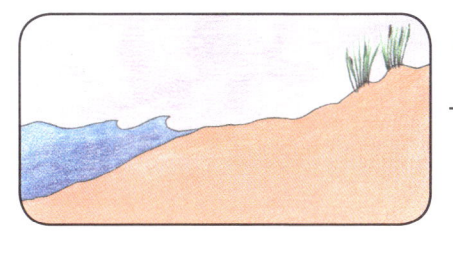

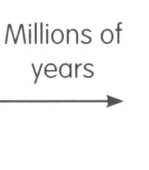

Smog

Air pollution that you can see.

Smog can form when ozone is present in the lower atmosphere. **Smog** can also form when sunlight is blocked by air pollutants given off by cars and factories.

Smoke

A hot cloud of tiny particles resulting from the burning of organic material.

Soil

A loose layer of weathered rock and organic matter that covers most of the Earth's land masses.

Soil is where most plants grow.

Earth and Space

Solar eclipse

A shadow cast on Earth when the Moon passes between the Earth and the Sun. If the Sun is not completely blocked, it is known as a *partial* **solar eclipse**. If the Sun is completely blocked, it is knows as a *total* **solar eclipse**.

Total Solar eclipse

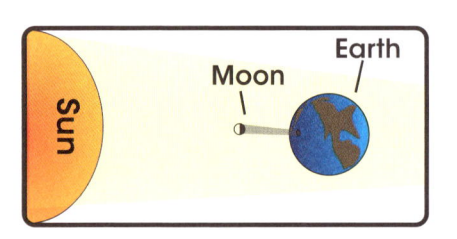

Solar system

Earth's planetary system. It contains one star, at least eight planets, moons, asteroids, and many other objects.

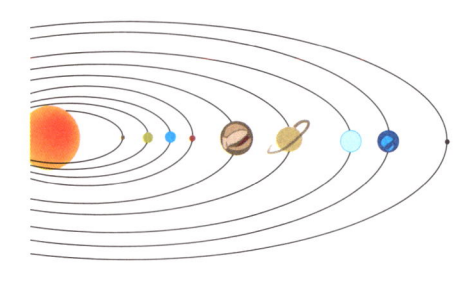

Space probe

A vehicle with no people on board that is used to explore space.

Scientists used **space probes** named *Spirit* and *Opportunity* to explore the planet Mars.

Star

A large ball of hot, glowing gas in space.

There are hundreds of billions of **stars** in the universe. The closest **star** to the Earth is the Sun.

Earth and Space

Subduction
The process by which one tectonic plate moves under another tectonic plate.

An area where **subduction** takes place is known as a *subduction zone*.

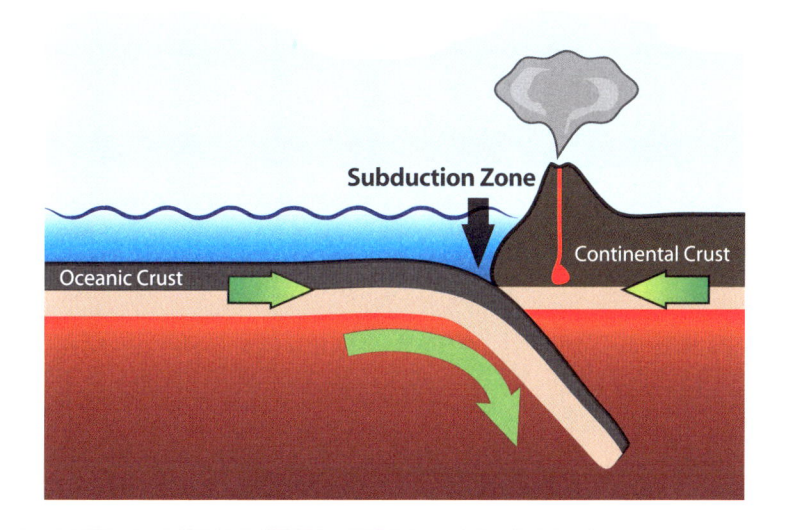

The prefix "sub-" means "under" or "below."

Summer solstice
The longest day of the year in the Northern hemisphere, and the shortest day of the year in the Southern hemisphere. The **summer solstice** usually falls on or around June 21st.

Surface water
Water that is located above the ground.

Water bodies such as oceans, seas, rivers, streams, wetlands, and lakes are all examples of **surface water**.

123

Earth and Space

Trench
A long, deep valley of the ocean floor that is associated with a subduction zone.

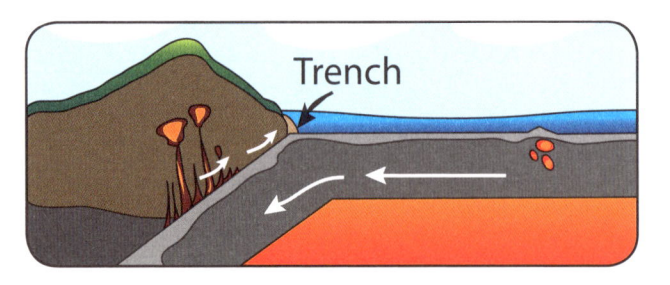

Universe
All planets, stars, and other objects that exist.

People explore the **universe** with space probes, shuttles, and telescopes.

Uplift
An increase in elevation of part of the Earth's crust.

Uplift is a result of the movement of the tectonic plates. When two plates collide with each other, one or both of the plates can be lifted up. **Uplift** can form mountain ranges.

Earth and Space

Vernal equinox
One of two days of the year when the Sun is directly overhead at the Equator (the other day is called the *autumnal equinox*). On this day, there are an equal number of hours of daytime and nighttime in both the Northern and Southern hemispheres.

Volcano
An opening on a planet's surface where lava flows out.

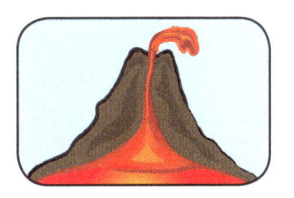

Waste
Anything made by people that they can't use.

Trash is **waste**. Hazardous **waste** is waste that is poisonous or dangerous.

Water cycle
The path that water takes as it moves out of the air as rain or snow, onto land or oceans, and then back into the air as evaporation.

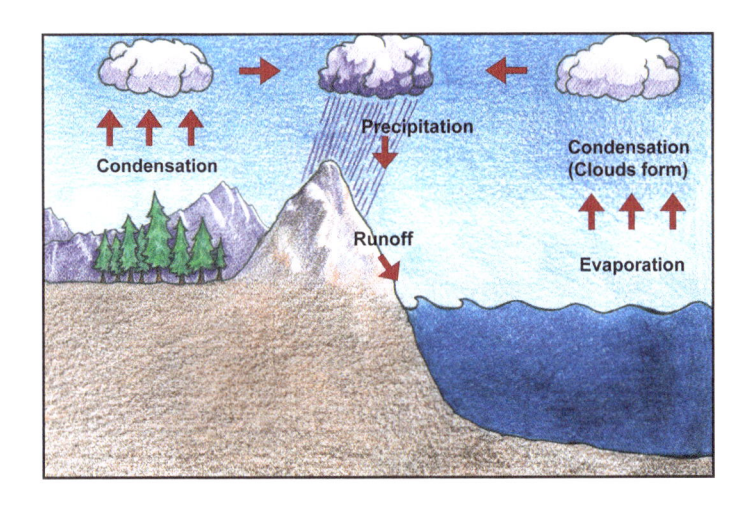

Earth and Space

Weather
The state of the atmosphere at a given time.

Weather includes temperature, precipitation, and wind. It is always changing.

Weathering
The breaking down of rocks by air, wind, water, or gravity.

Weather is a major factor in the **weathering** of rocks.

Winter solstice
The longest day of the year in the Southern hemisphere, and the shortest day of the year in the Northern hemisphere. The **winter solstice** usually falls on or around December 21.

The Periodic Table

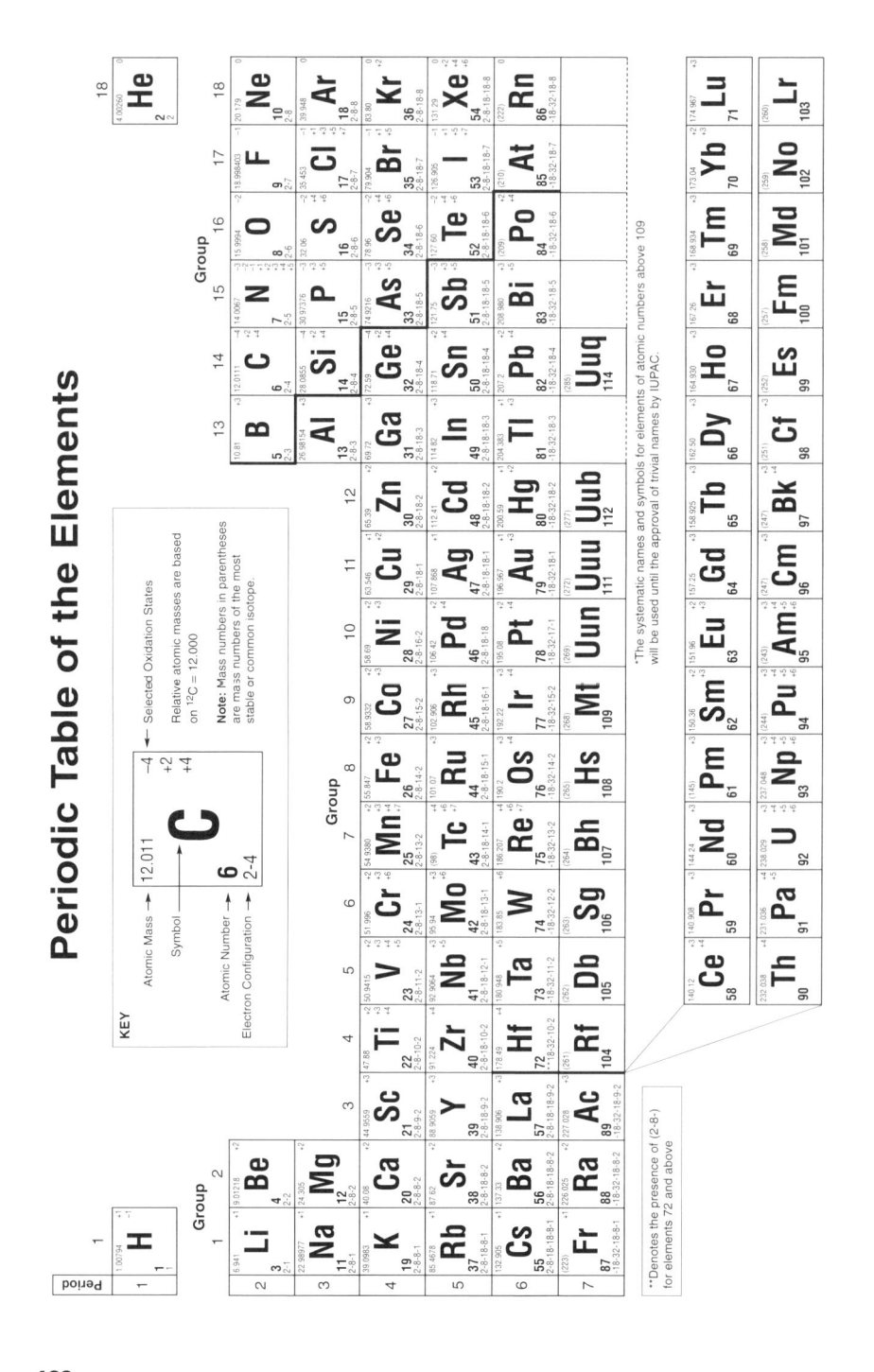

Periodic Table of the Elements

Index

A

Abiotic   12
Absolute zero   77
Absorb   77
AC circuit   77
Accelerate   77
Acid   53
Acid rain   99
Active transport   12
Adaptation   12
Allele   12
Alloy   77
Ampere   78
Amplitude   78
Anaphase   13
Animal   13
Antibiotic   13
Aquifer   99
Archaea   13
Asexual reproduction   14
Asteroid   99
Asthenosphere   99
Astronomical unit   100
Astronomy   100
Atmosphere   100
Atmospheric gases   100
Atmospheric pressure   53
Atom   53
Atomic mass unit   53
Atomic number   54
ATP   14
Autumnal equinox   100
Avogadro's law   54
Avogadro's number   54
Axis   78, 100

B

Bacteria   14
Barometer   54
Base   55
Behavior   14
Big Bang theory   101
Biodiversity   15
Biogenesis   15
Biological transport   15
Biosphere   15, 101
Biotic   16
Black hole   101
Boil   1
Boiling point   55

C

Camouflage   16
Cancer cells   16
Carbohydrate   16
Carbon   16, 55
Carbon cycle   17, 101
Catalyst   17, 55
Cell   18
Cell differentiation   18
Cell membrane   19
Cell wall   19
Centripetal force   79
Charge   79
Charged particle   56
Chemical   56
Chemical bond   56
Chemical change   56
Chemical formula   56
Chemical potential energy   57
Chemical property   57

Index

Chemical reaction   57
Chlorophyll   19
Chloroplast   20
Chromosome   20
Cilia   20
Circuit   79
Citric acid cycle   21
Classification   21
Cleavage   102
Climate   102
Clouds   102
Color   80
Combustion   1
Comet   102
Community   21
Compound   1
Conclusion   1
Condensation   2
Conductivity   57
Conservation of mass   58
Conserve   2
Constant   2
Consumer   21
Continent   103
Continental drift   103
Control   2
Convection   80, 103
Core   103
Coulomb   80
Covalent bond   58
Crest   80
Crust   104
Crystal   58
Current   81, 104

Cycle   2
Cytokinesis   22
Cytoplasm   22

D

Data   3
Daughter cells   22
DC circuit   81
Dead organic material   23
Decomposer   23
Delta   104
Density   3
Deposition   104
Diffusion   23
Diversity   24
DNA   24
Domain   24
Dominant   25
Drought   105

Index

E

Earthquake 105
Eclipse 105
Ecology 25
Ecosystem 25
Electric charge 81
Electric circuit 81
Electric conductor 81
Electric field 82
Electric insulator 82
Electricity 82
Electromagnetic radiation 83
Electron 59
Electron transport chain 25
Element 59
Embryo 25
Emission spectrum 83
Endangered species 26
Endoplasmic reticulum 26
Endothermic 59
Energy 3
Environment 26
Environmental influence 26
Enzyme 27
Equator 105
Equilibrium 3, 60
Erosion 106
Eukarya 27
Evaporation 4
Evidence 4
Evolution 27
Exothermic 60
External stimulus 27
Extinct 28

F

Fault 106
Fermentation 28
Fertile 28
Fertilization 28
Fission 83
Flagellum 28
Flood 106
Food chain 29
Food web 29
Force 84
Foreign substances 29
Fossil 29, 106
Fossil fuel 107
Fossil record 29
Freeze 4
Freezing point 60
Frequency 84
Freshwater 107
Friction 84
Fuel 107
Functional groups 60
Fungi 30
Fusion 84

G

Galaxy 107
Gamete 30
Gas 4
Gas laws 61
Gas volume 61
Gaseous 61
Gene 30
Gene expression 30

Index

Gene mutation   30
Gene pool   31
Genetic drift   31
Genetic engineering   31
Genetics   31
Genus   31
Geographic poles   107
Glaciation   108
Glacier   108
Global warming   108
Glucose   32
Glycolysis   32
Golgi complex   32
Gravitational force   85
Gravitational potential energy   85
Gravity   108
Greenhouse effect   109
Greenhouse gases   109
Groundwater   109
Grow   32

H

Habitat   33
Half-life   85, 109
Heat   85
Heat conductor   85
Heat flow   61
Heat insulator   86
Heredity   33
Hertzsprung-Russell Diagram   110
Hormone   33
Human being   33

Hydrogen   61
Hydrogen bond   62
Hypothesis   5

I

Ice age   110
Ideal gas   62
Igneous rock   110
Immune system   34
Independent assortment   34
Inhabit   34
Inherited characteristics   34
Inorganic   62
Interphase   35
Introduced species   35
Invasive species   35
Ion   62
Ionic bond   62

J

Joule   86

K

Kepler's laws   110
Kinematics   86
Kinetic energy   86
Kinetic molecular theory   63
Kingdom   36

Index

L

Lake   111
Land mass   111
Latent heat   63
Latitude   111
Lava   111
Lens   87
Life cycle   36
Light   5
Lipids   37
Liquid   5
Lithosphere   112
Longitude   112
Lunar eclipse   113
Lysosome   37

M

Macromolecule   37
Magma   113
Magnet   87
Magnetic field   87
Magnetic poles   88
Magnitude   5
Mantle   113
Marine organism   37
Mass   5
Matter   6
Mechanical energy   88
Melt   6
Melting point   63
Membrane   38
Mendel's laws of heredity   38
Metabolism   39
Metal   63

Metamorphic rock   113
Metamorphose   114
Metamorphosis   39
Metaphase   39
Meteor   114
Meteorite   114
Meteoroid   114
Microorganism   40
Migration   40
Mineral   115
Mitochondria   40
Mitosis   38, 40
Mixture   64
Model   6
Molarity   64
Mole   64
Molecule   6
Momentum   88
Moon   115
Motion   88
Mountain   115

N

Native species   41
Natural selection   41
Nebula   115
Neutron   65
Newton's Laws   89
Nitrogen   65
Nonliving   41
Nonmetal   65
Nonrenewable resource   7
Nuclear energy   66
Nuclear fission   66

133

Index

Nuclear fusion   66
Nuclear reaction   67
Nucleic acids   41
Nucleon   67
Nucleotide   41
Nucleus   42
Nutrient   42

O

Observe   7
Ocean   116
Offspring   42
One-dimensional collision   89
One-dimensional
    motion   89
Optical system   90
Orbit   116
Ore   116
Organ   43
Organ system   43
Organelle   43
Organic compounds   67
Organism   43
Oxidation number   67
Oxidation-reduction
    reaction   67
Oxygen   68
Ozone   116
Ozone depletion   117

P

Pangaea   117
Parallel circuit   90

Parent   43
Particle   7
Parts per billion   68
Parts per million   68
Passive transport   44
Pathogen   44
Percent composition   68
Period   69
Periodic Table   69
Periodic Table
    of the Elements   128
pH   69
Photosynthesis   44
Physical property   69
Planet   117
Planetary system   117
Plant   45
Plate   118
Plate tectonic theory   118
Polarization   90
Pollen   45
Pollination   45
Pollute   7
Pollution   7
Polymer   70
Population   45
Position   91
Potential energy   91
Power   91
Precipitate   70
Predator   46
Predict   7
Pressure   92

Index

Prey 46
Prime Meridian 118
Producer 46
Product 70
Prokaryote 46
Property 70
Prophase 47
Protein 47
Protist 47
Proton 71
Pure substance 71

Q

Qualitative data 71
Quantitative data 72

R

Radiation 72, 118
Radioactive decay 72, 119
Radiometric dating 72
Reactant 73
Recessive 47
Recycle 47
Redox reaction 73
Reflect 92
Refract 92
Renewable resource 8
Reproduction 48
Resistance 93
Resource 8
Respiration 48
Ribosome 48

RNA 48
Rock 119
Rock cycle 119
Rotation 120

S

Satellite 120
Saturated solution 73
Scalar 93
Sea-floor spreading 120
Sediment 120
Sedimentary rock 121
Segregation 48
Semiconductor 93
Series circuit 93
Sexual reproduction 49
Simple machine 93
Smog 121
Smoke 121
Soil 121
Solar eclipse 122
Solar system 122
Solid 8
Solubility 73
Solute 73
Solution 74
Solvent 74
Sound 94
Space probe 122
Species 49
Specific heat 74
Speed 8
Spindle 49

Index

Star   122
State of matter   74
Stationary charges   94
Stimulus   49
Subatomic molecules   75
Subduction   123
Substance   8
Summer solstice   123
Surface water   123

T

Taxonomy   50
Technology   9
Telescope   9
Telophase   50
Temperature   9
Tesla   94
Theory   9
Thermal energy   94
Thermodynamics   95
Thermometer   10
Three-dimensional motion   95
Time   10
Tissue   50
Titration   75
Torque   95
Trait   50
Trench   124
Trough   96
Two-dimensional motion   96

U

Uncontrolled cell division   51
Units   96
Universe   124
Unsaturated solution   75
Uplift   124

V

Vaccine   51
Vacuole   51
Valence electron   76
Vaporization   76
Variable   10
Vector   96
Velocity   10
Vernal equinox   125
Vibration   97
Virus   51
Viscosity   10
Volcano   125
Volt   97
Voltmeter   97
Volume   11

Index

W-Z

Waste   125
Waste materials   52
Water cycle   125
Watt   97
Wave   98
Wavelength   98
Weather   126
Weathering   126
Weight   11
Winter solstice   126
Work   98
Zygote   52

✳ Credits

Many thanks to the following organizations for the use of their photography and/or illustrations in this publication:

CDC Global Health Odyssey, photographers Judy Gantt and Mary Hilpertshauser

Centers for Disease Control and Prevention PHIL, photographer Janice Carr

HubbleSite (NASA) http://hubblesite.org

Integration and Application Network, University of Maryland Center for Environmental Science, ian.umces.edu/symbols

http://gallery.spacebar.org, photographer Tom Murphy VII

Marine Mammal Commission

National Oceanic and Atmospheric Administration www.noaa.gov

National Park Service, Rocky Mountain National Park

North Dakota State University Department of Soil Science www.soilsci.ndsu.nodak.edu

OSHA www.osha.gov

Rutgers University http://rwqp.rutgers.edu/reg_priority/nutrient_man/

Sue Nichols, Morgan State University

USDA Agricultural Research Service, photographer Jack Dykinga

US Fish and Wildlife Service

US Geologic Survey SOFIA (South Florida Information Access)

www.cepolina.com/freephoto

Additional Photo Credits
Capstone Press, cover, i (atom of water, H_2O); cover, i (DNA strand); cover, i (physics symbol) Ingram Publishing, cover, i (Saturn)